Multifunctional Metasurfaces

Design Principles and Device Realizations

Synthesis Lectures on Materials and Optics

Multifunctional Metasurfaces: Design Principles and Device Realizations

He-Xiu Xu, Shiwei Tang, Tong Cai, Shulin Sun, Qiong He, and Lei Zhou

ISBN: 978-3-031-01262-4   paperback
ISBN: 978-3-031-02390-3   ebook
ISBN: 978-3-031-00254-0   hardcover

DOI: 10.1007/978-3-031-02390-3

A Publication in the Springer series
SYNTHESIS LECTURES ON MATERIALS AND OPTICS

Lecture #5
Series ISSN
Synthesis Lectures on Materials and Optics
Print 2691-1930   Electronic 2691-1949

He-Xiu Xu#, Shiwei Tang#, and Lei Zhou*

#These authors contributed equally to this book.

*Corresponding author, email: phzhou@fudan.edu.cn

Multifunctional Metasurfaces

Design Principles and Device Realizations

He-Xiu Xu
Fudan University, China
Air Force Engineering University, China

Shiwei Tang
Fudan University, China
Ningbo University, China

Tong Cai
Fudan University, China
Air Force Engineering University, China

Shulin Sun
Fudan University, China

Qiong He
Fudan University, China

Lei Zhou
Fudan University, China

SYNTHESIS LECTURES ON MATERIALS AND OPTICS #5

ABSTRACT

In recent years, we have witnessed a rapid expansion of using super-thin metasurfaces to manipulate light or electromagnetic wave in a subwavelength scale. However, most designs are confined to a passive scheme and monofunctional operation, which hinders considerably the promising applications of the metasurfaces. Specifically, the tunable and multifunctional metasurfaces enable to facilitate switchable functionalities and multiple functionalities which are extremely essential and useful for integrated optics and microwaves, well alleviating aforementioned issues. In this book, we introduce our efforts in exploring the physics principles, design approaches, and numerical and experimental demonstrations on the fascinating functionalities realized. We start by introducing in Chapter 2 the "merging" scheme in constructing multi-functional metadevices, paying particular attention to its shortcomings issues. Having understood the merits and disadvantages of the "merging" scheme, we then introduce in Chapter 3 another approach to realize bifunctional metadevices under linearly polarized excitations, working in both reflection and transmission geometries or even in the full space. As a step further, we summarizes our efforts in Chapter 4 on making multifunctional devices under circularly polarized excitations, again including designing principles and devices fabrications/characterizations. Starting from Chapter 5, we turn to introduce our efforts on using the "active" scheme to construct multifunctional metadevices under linearly polarized wave operation. Chapter 6 further concentrates on how to employ the tunable strategy to achieve helicity/frequency controls of the circularly polarized waves in reflection geometry. We finally conclude this book in Chapter 7 by presenting our perspectives on future directions of metasurfaces and metadevices.

KEYWORDS

multifunctional metasurface, polarization-dependent, helicity control, tunable, full-space, geometric phase, propagation phase, reflection, transmission

Contents

CHAPTER 1

Introduction

Manipulating electromagnetic (EM) waves in the desired manners is one of the most important tasks in photonics research, as photons are believed to be alternative information carriers that may play crucial roles in the next-generation industry revolution. According to Maxwell's equations, permittivity and permeability of a medium dictate the behaviors of EM waves propagating inside it. However, naturally existing materials only exhibit limited abilities to control EM waves, since their permittivity ε lie in a narrow variation range and even worse, their permeability μ are all very close to 1 at high frequencies (e.g., the visible range), due ultimately to the weak interactions between natural molecules/atoms and magnetic fields of EM waves. As a result, typically conventional EM devices need to be thick enough (compared to the operation wavelength) and exhibit certain curved shapes to ensure appropriate propagation phases accumulated, to realize the desired wave-manipulation functionalities (say, focusing). Besides, efficiency is also an issue for conventional devices caused by an impedance mismatch between air and natural materials which typically do not exhibit magnetic responses. Such limitations (i.e., size, shape, and efficiency) significantly hinder the applications of conventional optical devices in next-generation on-chip photonics scenarios which are typically flat, ultra-compact, and energy saving. Meanwhile, facing the increasing demands on data-storage capacity and information processing speed in modern science and technology, EM integration plays a more and more important role and has attracted intensive attention with remarkable applications. An ultimate goal pursued by scientists and engineers along this development is to make miniaturized devices as small as possible yet equipped with powerful functionalities as many as possible. However, available efforts based on conventional materials suffer from the issues of device thickness, low efficiency, and restricted functionalities.

Metamaterials (MTMs) [1, 2], consisting of deep-subwavelength-sized EM microstructures (e.g., meta-atoms) arranged in periodic or non-periodic orders, have drawn much attention recently. Through tailoring the microstructures of meta-atoms, MTMs can in principle exhibit arbitrary values of ε and μ, thus exhibiting extraordinarily strong capabilities to control EM waves. Many fascinating wave-manipulation effects have been demonstrated based on MTMs, such as negative refraction [3, 4], super-resolution imaging [5, 6], cloaking [7–9], polarization-control [10–13], perfect light absorption [14], and transparency [15, 16], and unusual wave-control effects realized by zero-index MTMs [17, 18]. Attempts have also been made to achieve multifunctional EM devices based on MTMs. However, the realized devices typically exhibit bulky sizes and low efficiencies, since MTMs are three-dimensional (3D) materials composed by resonant metallic structures which can easily absorb EM waves. Moreover, such 3D devices require complex fabrication processes, adding more disadvantages to EM integrations [19, 20].

Metasurfaces, ultrathin MTM layers constructed by planar meta-atoms of pre-determined EM responses arranged in specific two-dimensional (2D) orders, can largely overcome the difficulties faced by MTMs. In recent years, we have witnessed a rapid development of using ultrathin metasurfaces to manipulate light or EM waves on deep-subwavelength scales. Tuning the EM responses of meta-atoms to realize certain transmission/reflection phase distributions on the metasurfaces, one can use these ultra-thin devices to efficiently reshape the wavefronts of incident EM beams based on Huygens' principle, achieving unusual effects including anomalous beam bending based on generalized Snell's law [21–29], propagating wave to surface waves conversion [30–32], polarization-control [33–39], focusing [40–42], holograms [43–45], flat-lens imaging [46–50], tunable devices [51–53], photonic spin-Hall effect [54–56], etc. In contrast to MTM-based EM devices replying critically on the propagation phases of EM waves, metasurfaces fully exploited the abrupt phase changes of EM waves at the meta-atom interfaces, and therefore, they can be much thinner than the working wavelength. Meanwhile, since typically EM waves do not stay inside metallic structures for a long time, the loss issue can be significantly alleviated in metasurface-based devices. Also, the flat configuration makes such systems easy to fabricate. Most importantly, by tailoring the "order" on which the meta-atoms are arranged, one can realize metasurfaces exhibiting desired inhomogeneous distributions of amplitude and phase for transmitted/reflected EM waves, enabling diversified fascinating wave-manipulation effects far beyond those realized with MTM-based devices.

These unique features make metasurfaces ideal candidates to realize flat and miniaturized metadevices exhibiting powerful wave-manipulation capabilities, being particularly suitable for on-chip photonic applications. Typically, the devices made by metasurfaces are flat, much thinner than the wavelength, and exhibit much higher efficiencies than their bulky MTM counterparts, all being highly favorable for integration-optics applications. These attractive properties make metasurfaces the best candidates to construct multifunctional EM devices. Indeed, many efforts have recently been devoted to designing multifunctional optical devices based on metasurfaces, typically using polarization, helicity, frequency, or incidence angle of the impinging light as external knobs to control the functionalities exhibited by the devices. The proposed/demonstrated devices are usually equipped with functionalities combining two or more of those demonstrated before on single-function metasurfaces, such as beam bending, focusing, hologram, surface-wave conversion, directive beaming, etc. In what follows, we briefly summarize the available mechanisms so far developed to realize multifunctional metadevices.

1. Polarization is one important degree of freedom that can be used to realize multifunctional metadevices. Based on anisotropic meta-atoms with polarization-sensitive EM responses, various metadevices were realized at different frequencies, exhibiting multiple functionalities triggered by incident EM waves with different linear polarizations (LPs) [57–63]. In 2014, a dual-functional meta-hologram was experimentally demonstrated based on

anisotropic metal–insulator–metal (MIM) meta-atoms [64]. However, the device still exhibits similar functionalities (e.g., hologram) for two incident polarizations. In 2016, a general strategy was proposed to design bifunctional metadevices exhibiting distinct functionalities with very high efficiencies [57]. Very recently, such a concept was successfully implemented in the optical regime [59]. For transmissive metadevices, in addition to using multilayer meta-atoms, Huygens' meta-atoms and dielectric meta-atoms were also frequently adopted to construct multifunctional metadevices with distinct functionalities in the near-infra-red (IR) regime [65].

2. Helicity is another degree of freedom frequently exploited to design multifunctional metadevices based on the Pancharatnam–Berry (PB) mechanism [66–73]. A commonly applied scheme is to merge several different PB metasurfaces, each exhibiting a certain functionality as the incident circular-polarized (CP) light takes a particular helicity, into one single device [70, 71]. Although such a "merging" strategy is physically straightforward, realized devices exhibit limited working efficiencies and suffer from the issue of functionality cross-talking. In 2015, PB metasurfaces were proposed to record various hologram patterns into helicity-multiplexing channels [66]. While the realized hologram exhibits broadband properties, the measured working efficiency was very low. Recently, Hasman's group experimentally demonstrated that the alliance of spin-enabled geometric phase and shared-aperture concepts can open a new pathway to implement photonic spin-controlled multifunctional metasurfaces [73, 74]. Helicity-controlled multiple wavefronts such as vortex beams were demonstrated in the visible regime [67]. Combining geometric and propagating phases, chiral holograms were experimentally demonstrated based on transmissive dielectric metasurfaces, which can efficiently generate independent far-field images for right circular polarization (RCP) and left circular polarization (LCP) excitations [68].

3. Wavelength-multiplexing is also widely exploited to realize multifunctional metasurfaces. In 2016, multicolor meta-holography was experimentally realized with a single type of plasmonic pixel, based on an off-axis illumination method [75]. In a parallel line, dielectric metasurfaces were used to build high-efficiency wavelength-multiplexing metadevices at optical frequencies, based upon carefully designed silicon nano blocks [76]. Again, the experimentally achieved efficiencies of the reconstructed images for highly dispersive color holograms are limited due to the intrinsic issue of the "merging" concept.

4. Incidence angle was recently identified as another degree of freedom to be exploited. In 2017, Kamali et al. proposed an angle-multiplexed metasurface, composed of reflective high dielectric contrast U-shaped meta-atoms with incidence-angle-sensitive responses, and experimentally demonstrated that it can realize high-efficiency angle-multiplexed diffractions and holograms at the working wavelength of 915 nm [77]. Zhou's group established a theory to quantitatively describe the angular dispersion in metasurfaces [78] and

propose a general strategy to realize angle multiplexed metadevices [79], which can exhibit distinct wave-front-control functionalities as shined at different incident angles. Such angle multiplexed metadevices are realized through carefully controlling both the near-field couplings between meta-atoms and the radiation pattern of a single meta-atom.

5. Adding "active" elements into metasurfaces is another important approach to achieve multifunctional metadevices, with functionalities controlled by external stimuli. Such an "active" scheme stimulates a lot of research works recently, forming a very "active" research sub-field. Below we briefly mention several representative approaches, categorized based on the external stimuli used.

5.1 **Electrically sensitive materials** include varactor diodes, liquid crystals, doped semiconductors, and 2D materials, functioning in different frequency regimes. At microwave frequencies, varactor and PIN diodes are frequently adopted in designing tunable metasurfaces [80–84], since their EM responses (e.g., capacitances) can be dramatically tuned by applying external voltages. In THz and IR regimes, doped semiconductors are often used, since their conductivities (and thus optical responses) can be dramatically modified via electrical gating. Such a mechanism features broad bandwidth and high modulation speed and is compatible with C-MOS technology [85–91]. Graphene, a zero-bandgap 2D material with conductivity tuned efficiently via external gating, is another excellent candidate to help realize tunable devices in the THz or mid-IR regimes, typically in combination with carefully designed metasurfaces [92–107]. Liquid crystals are also widely used as "active elements," since the orientation angles of molecules such media can be controlled by external electric fields, leading to significant modulations on the refractive index. Resonant properties of metasurfaces with liquid crystals incorporated can thus be efficiently tuned by applying the electric field across the liquid crystals [108–110].

5.2 **Optically sensitive materials (OTMs)** can be hybridized with passive metasurfaces to realize tunable multifunctional metadevices. OTMs include semiconducting materials and optically responsible phase-change materials, whose optical properties strongly depend on the pumping light. Combining these OTMs with carefully designed meta-atoms, various functionality-tunable metadevices were realized by different groups in different frequency regimes [111–117].

5.3 Exerting **mechanical forces** on a metasurface can also efficiently tune its optical properties since the meta-atom structure or its local environment can be dramatically modified. Micro-electro-mechanical-system (MEMS) or nano-electro-mechanical-system (NEMS) are widely used technologies to realize mechanically tunable metasurfaces in different frequency domains [118–124]. Mechanically tunable metasurfaces based on stretchable and flexible substrates have been proposed to achieve various dynamic effects (e.g., color tuning [125, 126], switchable holograms [127], and varifocal lenses [128]).

5.4 Incorporating **thermally responsive materials** into metasurfaces is another approach to realizing tunable metasurfaces. Vanadium dioxide (VO2) [129, 130], liquid crystals, and superconductors [131, 132]) are typical examples of this class, which exhibit temperature-dependent optical properties. Therefore, these materials were also frequently adopted to design tunable metadevices exhibiting different functionalities [129–135].

5.5 In addition to the mechanisms introduced above, many others have also been successfully used to realize tunable metadevices in different frequency ranges. For example, one can incorporate ferroelectric materials into the meta-atom design, which can be tuned by applying a magnetic field [136]. Liu et al. demonstrated a kind of dynamic plasmonic color display technology that can realize tunable metadevices based on a chemical approach [137]. Controlled hydrogenation and dehydrogenation of the constituent magnesium nanoparticles, which serve as dynamic pixels, allow for plasmonic color printing, tuning, erasing, and restoration of color.

Having briefly mentioned existing approaches in making multifunctional metadevices, in this book, we would like to summarize our efforts devoted to this fast-developing field in the past several years. To benefit our readers, we will introduce the physics principles and design approaches of these metadevices, in addition to presenting the numerical and experimental demonstrations on the fascinating functionalities realized. This book is organized in the following way. We start by introducing in Chapter 2 the "merging" scheme developed in the early years and widely adopted by several groups in constructing multi-functional metadevices, paying particular attention to its shortcomings issues (e.g., low operating efficiencies and functionality cross-talking). Having understood the merits and disadvantages of the "merging" scheme, we then introduce in Chapter 3 another approach to realize bifunctional metadevices under LP excitations, working in both reflection and transmission geometries or even in the full space. In addition to presenting several high-efficiency metadevices experimentally demonstrated with various combinations of functionalities, we focus on the working principle and design approach to achieve such devices, including how to diminish the cross-talkings between different functionalities. Chapter 4 summarizes our efforts on making multifunctional devices under CP excitations, again including designing principles and devices fabrications/characterizations. Starting from Chapter 5, we turn to introduce our efforts on using the "active" scheme to construct multifunctional metadevices. Specifically, we first introduce in Chapter 5 the design principles to realize such tunable metadevices under LP excitations and then present several proof-of-concept demonstrations of metadevices in both reflection and transmission geometries exhibiting switchable wave-control functionalities. Next, concentrate on how to employ the tunable strategy to achieve helicity/frequency controls of the CP waves in reflection geometry (Chapter 6). We finally conclude the book in Chapter 7 by presenting our perspectives on future directions of metasurfaces and metadevices.

CHAPTER 2

Early Attempts on Multifunctional Metasurfaces: The "Merging" Concept

2.1 DESIGN PRINCIPLES

Facing increasing demands on speed and memory of EM devices, EM integration is highly desired in modern science and technology. Metasurfaces are ideal candidates to integrate multiple diversified functionalities into single devices with deep-subwavelength thickness and high efficiencies [138]. Various approaches have recently been proposed to achieve this goal, and a simple and physically straightforward scheme developed in the early years utilized the so-called "merged" meta-structures to achieve multifunctional metasurfaces. In such a scheme, people first design individual metasurfaces exhibiting their functions (e.g., one for holographic image and one for vortex beam) and then construct a multifunctional device simply through merging the two structures. Below we present several examples to illustrate how the scheme works.

In order to introduce the "merging" concept more clearly, we take a typical bifunctional metasurface based on the "merging" concept as an example. Figure 2.1 presents an optical bifunctional metasurface that can realize a hologram image or a vortex beam, depending on the helicity of excitation light [139]. To achieve their end, the authors first design two individual metasurfaces (both utilizing the metal-bar structure as basic meta-atoms) which can realize one of the needed functionalities when they are shined by incident light taking circular polarization (CP) with different helicities (see Fig. 2.1). The desired phase profiles on two metasurfaces are created by the PB principle [54, 140] through rotating the metallic bars at different positions by appropriate angles. Since the two metasurfaces exhibit identical periodic structures and there are enough open spaces between metallic bars, the authors then merge two metasurfaces to obtain the final design in which all metallic bars do not touch with each other. Such a device was finally fabricated out and experimentally characterized, showing nice bifunctional performances (Fig. 2.1). However, the working efficiency of the device is quite low, which is found to be around 9% [139].

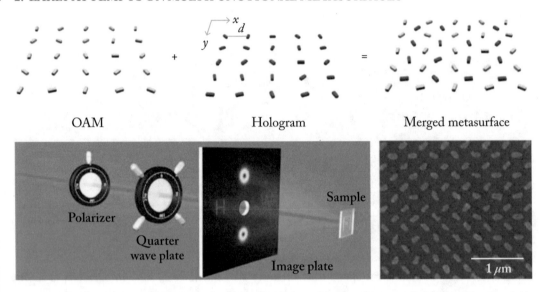

Figure 2.1: **Multifunctional device designed with merged structures.** Design strategy, sample picture, and experimental characterizations of a multifunctional metasurface than can generate holographic images or a vortex beam depending on the helicity of incident circularly polarized light.

2.2 REFLECTION-GEOMETRY REALIZATIONS

Having understood the key issues in the "merging" concept, based on which people designed kinds of multifunctional metasurfaces and first is in reflection geometry because high-efficiency reflective meta-atoms are much easier to find than their transmissive counterparts.

Wen et al. further experimentally demonstrated a design methodology to achieve helicity multiplexed functionalities by combining two sets of hologram patterns operating with opposite incident helicities on the same metasurface [141]. First, the Gerchberg–Saxton algorithm is used to generate two-phase profiles, which can reconstruct two off-axis images on the different sides of the incident light. Both images ("bee" and "flower") have a projection angle of $22° \times 22°$ and an off-axis angle of $10.35°$ in the imaging area. Then, the two-phase profiles are encoded onto the metasurfaces (Fig. 2.2a), where the nth phase pixel φ_n of the hologram is represented by a nanorod with the orientation angle $\varphi_n/2$ defined in the metasurface. After that, two sets of data are merged with a displacement vector of $(d/2, d/2)$, as shown in Fig. 2.2a. d is the distance between neighboring antennas with a value of 424 nm. Therefore, although the new metasurface contains two sets of hologram data, the size of the sample is still the same and the equivalent pixel size is 300×300 nm, leading to an increase of the nanoantenna density. On the illumination of left circular polarization (LCP) light, the merged metasurface can reconstruct the 'flower' on the left and the "bee" on the right side of the metasurface viewing from the incident

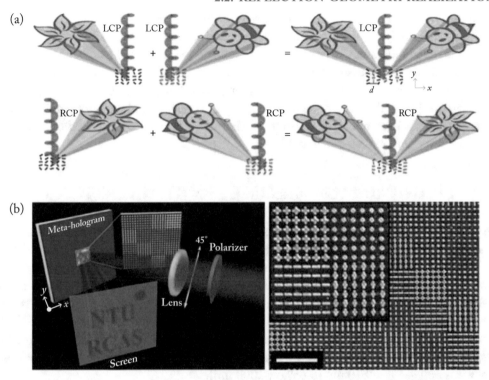

Figure 2.2: (a) A metasurface that can generate multiple hologram images as shined by circularly polarized light with different helicity. (b) Schematic and scanning electron microscope image of the metahologram for projecting the polarization-controlled dual images of "NTU" and "RCAS."

beam, respectively (Fig. 2.2a). Since the sign of the phase profile can be flipped by controlling the helicity of the incident light, the positions of reconstructed images in Fig. 2.2a are swapped in contrast to those in Fig. 2.2a when the helicity of the incident light is changed from LCP to right circular polarization (RCP).

As a result, the reconstructed images on the same position are switchable, that is, either "bee" or "flower," depending on the helicity of the incident light.

Tsai's group presented a reflection-type metahologram for visible light that can overcome the issues mentioned above [64]. The basic meta-atoms are MIM structures with the top resonators being gold nanocrosses, which inherently possess much higher efficiencies than the V-antennas proposed by Shalaev's group [142]. By tuning the structural details, one can easily obtain a set of meta-atoms that yield different reflection phases to cover the full 2π range, yet with reflection, amplitudes staying at high values. Moreover, the lengths of two orthogonal nanorods are free parameters to independently control the reflection phases for two LPs, which allowed the authors to design a single metasurface encoding two distinct hologram images (i.e., "NTU" and "RCAS") when illuminated by light with different polarizations (see Fig. 2.2b). The realized metaholograms can work within broadband (width $\cong$ 880 nm) under a wide range of incident angles. The measured efficiency of the device is about 18% at the wavelength of 780 nm, while the simulated value is even higher (28%). Compared to the transmission-type metaholograms [142], this scheme yields much higher efficiency and does not require polarization conversion for detection. The working efficiency can be further improved by introducing more phase levels and reducing material losses.

2.3   TRANSMISSION-GEOMETRY REALIZATIONS

Such a "merging" concept has been straightforwardly applied to realize many other transmission-geometry multifunctional metadevices.

Similar to the reflection-geometry realizations mentioned above, a commonly applied scheme is to merge several different PB metasurfaces, each exhibiting a certain functionality as the incident CP light takes a particular helicity, into one single device [71], which is shown in Fig. 2.3a. In 2015, Huang et al. proposed to use PB metasurfaces to record various hologram patterns into helicity-multiplexing channels, as shown in Fig. 2.3c [66]. Although a broadband hologram ranging from 633–1000 nm was experimentally achieved, the measured working efficiency was only 4.5% at 810 nm (0.65% at 1000 nm).

The transmission-geometry multifunctional metadevices can work tools for polarization imaging and image processing [143, 144]. Chen's group experimentally demonstrated a light sword metasurface lens with multiple functionalities. As shown in Fig. 2.3b, the position of focal segments can be controlled by changing the polarization state of the incident light [143]. To design such a multifunctional device, two metasurfaces (each one for a specific focal segment) are designed to operate with opposite incident helicities and merged with a displacement. The conversion efficiency between the polarization states based on the plasmonic metasurface is measured to be 2% at 650 nm, which is at the lower edge of what is required for practical applications.

Yang's group experimentally demonstrate chiral geometric metasurfaces based on intrinsically chiral plasmonic stepped nanoapertures with a simultaneously high circular dichroism in transmission and large cross-polarization ratio in transmitted light to exhibit spin-controlled wavefront shaping capabilities [145]. As shown in Fig. 2.3d, the chiral geometric metasurfaces are constructed by merging two independently designed subarrays of the two enantiomers for the stepped nanoaperture. Under a certain incident handedness, the transmission from one subarray is allowed, while the transmission from the other subarray is strongly prohibited. The merged metasurface then only exhibits the transmitted signal with the phase profile of one subarray, which can be switched by changing the incident handedness. Based on the chiral geometric metasurface, both chiral metasurface holograms and the spin-dependent generation of hybrid-

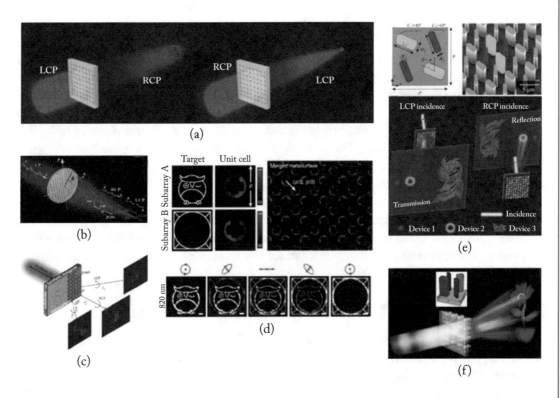

Figure 2.3: (a) Schematic to show the polarization selectivity of the metasurface device. The metasurface functions as a hologram to reconstruct an image of "cat" or a lens that converges the incident light into the focal point, depending on the incident/detected light combination. (b) Schematic of multifunctional light sword metasurface lens. Upon the illumination of incident light with linear polarization (LP), the device has two real focal segments corresponding to the transmitted light with RCP and LCP. (c) Schematic illustration of the hybrid multiplexing holograph based on PB Metasurfaces. (d) Illustration of the chiral metasurface hologram merging subarray A and subarray B to enable spin-controlled wavefront shaping. Subarrays A and B are contributing to the images of "owl" and "window," respectively. Reconstructed images for the chiral metasurface hologram at 820 nm for different polarization states. (e) 3D views of all-silicon supercells and scanning electron microscope image of the fabricated metasurface. The schematic illustration of the designed devices that generate diffraction patterns in the transmission (or reflection) field under the illumination of LCP (or RCP). (f) Dielectric wavelength-multiplexed metasurfaces for achromatic and dispersive holograms.

order Poincaré sphere beams are experimentally realized. When the chiral geometric metasurface is switched on, the holography efficiency is measured to be 6.8% at 820 nm.

As an alternative to metallic nanorods, however, a dielectric metasurface can be used to dramatically increase this value since it can decrease the ohmic losses and improve the scattering cross-sections of the metal nano-rods. An all-dielectric monolayer metasurface is proposed by Luo's group to simultaneously realize circular asymmetric transmission (AT) and wavefront shaping based on asymmetric spin-orbit interactions (Fig. 2.3e) [146]. Circularly polarized incidence, accompanied by arbitrary wavefront modulation, experiences spin-selected destructive or constructive interference. An extinction ratio of $\approx 10 : 1$ and an AT parameter of ≈ 0.69 at 9.6 μm, as well as a full-width half-maximum of ≈ 2.9 μm ($\approx 30\%$ of the peak wavelength), is measured with the designed metasurface. As far as it is known, this is the first report on the realization of simultaneous giant AT and arbitrary wavefront modulation with only one metasurface. In a parallel line, other dielectric metasurfaces are used to build high-efficiency wavelength-multiplexing metadevices at optical frequencies. As shown in Fig. 2.3f, Wang et al. experimentally demonstrated that a metasurface formed by three kinds of silicon nano blocks multiplexed in a subwavelength super-unit can achieve wavefront manipulations for red, green, and blue light, simultaneously [76].

2.4   ISSUES WITH THE "MERGING" SCHEME

In reviewing these metadevices based on the "merging" concept, we find that the proposed design strategy is physically transparent and easy to implement. However, to make the "merging" process work, the adopted meta-atoms must be very simple structures (say, metal bar) to avoid metallic overlapping. Unfortunately, these meta-atoms typically do not satisfy the 100%-efficiency criterion established for PB metasurfaces [54], and thus one type of meta-atoms can generate background noises in addition to the desired functionalities. As a result, such metadevices typically suffer from the issues of low operating efficiencies and functionality cross-talking, except [141] where the issue was partially solved by seeking a high-efficiency PB meta-atom in reflection geometry. Although the dielectric metasurface can decrease the ohmic losses and improve the scattering cross-sections [76, 146], the experimentally achieved working efficiencies of the are also limited due to the intrinsic issue of the "merging" concept. In conclusion, such a "merging" strategy is physically straightforward but the realized devices exhibit limited working efficiencies and suffer from the issue of functionality cross-talking.

CHAPTER 3

Multifunctional Metasurfaces/Metadevices Based on Single-Structure Meta-Atoms I: Linear-Polarization Excitations

While the "merging" scheme described in the last section is simple and easy to implement, it faces several severe issues that limit its practical applications, as discussed at the end of the last section. In this chapter, we propose an alternative strategy to design high-efficiency bifunctional metasurfaces in responses to EM waves with LPs [57, 58, 61], based on single-structure meta-atoms with anisotropic EM properties. We first present the working principles of such high-efficiency bi-functional metasurfaces working in reflection, transmission, and even full-space geometries, respectively. Based on the derived criteria for different cases, we next introduce the designs and fabrications of several bifunctional metadevices and employ microwave experiments, including far-field and near-field properties, to demonstrate their pre-designed bi-functionalities.

3.1 DESIGN PRINCIPLES: EFFICIENCY AND POLARIZATION CROSS-TALK

We first present our design principle to realize multi-functional metadevices/metasurfaces with high efficiencies [57, 58]. In this chapter, the multifunctional metasurfaces are excited in linear-polarizations, and the system exhibits global mirror symmetries with respect to $x \to -x$ and $y \to -y$ operations. The EM characteristics of those meta-atoms can be described by two diagonal Jones' matrices:

$$\boldsymbol{R}(x, y) = \begin{pmatrix} r_{xx}(x, y) & 0 \\ 0 & r_{yy}(x, y) \end{pmatrix}$$

and

$$T(x, y) = \begin{pmatrix} t_{xx}(x, y) & 0 \\ 0 & t_{yy}(x, y) \end{pmatrix},$$

with r_{xx}, r_{yy}, t_{xx}, t_{yy} being the reflection/transmission coefficients for waves polarized along two principal axes $\hat{x}$ and $\hat{y}$ for each meta-atom. To achieve high efficiency for the metasurface working in reflection geometry, we need a reflective metasurface with $T \equiv 0$ and $|r_{ii}(x, y)| = 1$. For transmission geometry, a totally transmissive meta-atoms with $R \equiv 0$ and $|t_{ii}(x, y)| = 1$ should satisfy. While in order to realize high-efficiency metasurfaces working in full-space conditions, we need our meta-atoms to be perfectly reflective and transparent for two orthotropic polarizations, such as totally reflective for an $\hat{x}$-polarized incidence and perfectly transparent for a $\hat{y}$-polarized case. Moreover, in all cases, the corresponding phases with non-zero coefficients should be freely tuned within a 2π variation range by varying the structural parameters of the meta-atoms. If different meta-atoms with desired phase distributions (i.e., $\phi_{xx}^{r/t}(x, y)$ and $\phi_{yy}^{r/t}(x, y)$ for reflection or transmission geometry, $\phi_{xx}^{r}(x, y)$ and $\phi_{yy}^{t}(x, y)$ for full-space metasurfaces) can be optimized, we can thus design metadevices/metasurfaces with pre-determined bifunctionalities with very high efficiencies.

We then describe another key factor to design multi-functional metasurfaces by suppressing the polarization crosstalk [61]. For an arbitrary anisotropic metasurface under the Cartesian coordinate system, we require four variables to describe the phase gradients $\xi_x(x), \xi_y(x), \xi_x(y)$, and $\xi_y(y)$ along different directions under different polarizations,

$$\begin{bmatrix} \xi_x(x) & \xi_y(x) \\ \xi_x(y) & \xi_y(y) \end{bmatrix} = \begin{bmatrix} \dfrac{\delta\varphi(x, y)}{\delta x} & \dfrac{\delta\varphi_y(x, y)}{\delta y} \\ \dfrac{\delta\varphi_x(x, y)}{\delta x} & \dfrac{\delta\varphi_y(x, y)}{\delta y} \end{bmatrix}. \tag{3.1}$$

Here, x or y in the subscript of $\xi_x(x), \xi_y(x), \xi_x(y)$, and $\xi_y(y)$ denotes the gradient direction, whereas that in the bracket represents the polarization direction. We thus have four degrees of freedom to describe and design anisotropic metasurface. According to the generalized Snell's law, the transverse (k_x and k_y) and longitudinal (k_z) wave vectors of the EM wave scattered by the metasurface (shined by a normally incident EM wave) are given by

$$\begin{cases} k_x(x/y) = \xi_x(x/y) \\ k_y(x/y) = \xi_y(x/y) \\ k_z(x/y) = \sqrt{k_0^2 - k_x^2(x/y) - k_y^2(x/y)}. \end{cases} \tag{3.2}$$

Equation (3.2) shows that the wave-vector of the scattered EM wave can be controlled by both the incident polarization and the related phase gradient. These expanded freedoms are extremely helpful to achieve diversified EM characteristics. In practice, one typically obtains the

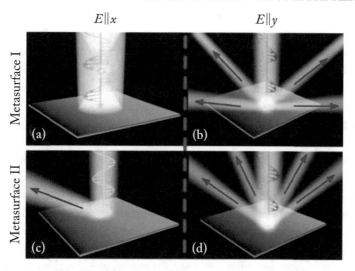

Figure 3.1: Schematic illustration and working principle of two flexible emission systems based on two bifunctional metasurfaces. (a, b) For emission system I, the metasurface functions as (a) a lens to focus the beam and (b) a beam splitter to generate quad large-angle pencil beams with polarizations $E\|x$ and $E\|y$, respectively. (c, d) For emission system II, the metasurface functions as (c) a beam deflector to steering the beam and (d) a beam splitter to generate quad small-angle pencil beams with polarizations $E\|x$ and $E\|y$, respectively.

desired $\xi_x(\sigma)$ and $\xi_y(\sigma)$ properties through changing the geometrical parameters of the meta-atom which are sensitive to the σ-polarized EM wave. However, for many meta-atoms with polarization cross-talk, the variation of structural parameters along the x-direction often strongly affects the phase gradients/profiles related to the y-polarization, and thus making it hard (if not impossible) to realize an accurate and fast design for the four gradients simultaneously.

3.2   REFLECTION-GEOMETRY REALIZATIONS: VERSATILE BEAM-CONTROL METADEVICES

The versatile beam-control metadevices that we designed are shown in Fig. 3.1. Two proof-of-concept metadevices, each combining two distinct complex functionalities including steered-beam, focused-beam, and multi-beam directional emissions. In both cases, the complicated phase profiles for two distinct functionalities require 2D geometrical-parameter searching and thus our strategy can save lots of time in designing such devices. The flexibly controlled highly directive emissions with multifunctionalities have huge fascinations and prospects to conveniently integrate complex systems with low costs.

Here, we propose a strategy to exhibit nearly negligible polarization cross-talk, such that the four phase gradients can be independently designed. The proposed anisotropic meta-atom contains two identical composite metallic resonators and a continuous metal plate separated by two dielectric spacers (2.5 mm–thick F4B board with $\varepsilon_r = 2.65 + 0.001i$); see Fig. 3.2a. To diminish the polarization cross-talk and broaden the bandwidth, we purposely designed the metallic resonator to contain both a metallic cross and an external wire loop. These two structures resonate at two different frequencies. Since the meta-atom is electrically subwavelength (0.277λ at 10 GHz), the two resonant modes can be quantitatively described by an equivalent circuit model (CM) shown in Fig. 3.2b, where the lower and upper mode around f_1 and f_2 are physically modeled by a series resonant tank formed by L_1, C_1, R_1, and L_2, C_2, R_2, respectively. Here, L, C, and R represent the effective inductance, capacitance, and resistance (absorption) of the circuit. Figure 3.2c compares the spectra of reflection phase and amplitude for the single-layer and double-layer meta-atoms based on finite-difference-time-domain (FDTD) calculations. In the single-layer case, two reflection dips are clearly observed around $f_1 = 6.12$ and $f_2 = 9.9$ GHz resulted from two magnetic resonant modes generated by the couplings between the cross and its surrounding loop with the metallic ground plane, evidenced by the reversed currents on metallic patterns and ground plane as shown in Fig. 3.2g. Cascading the two resonances appropriately can significantly enhance the working bandwidth. Full-wave simulations (lines) are in good agreement with the CM calculations (symbols). When adding another resonator to form a double-layer meta-atom, couplings between different layers split the two resonances into four (dashed lines in Fig. 3.2c), and thus the working bandwidth is further enhanced, providing us more freedoms to engineer the phase slope. As depicted in Fig. 3.2d, changing l_y from 0.5–3.65 mm leads to a large phase variation φ_y of near 380° (full 360° cover), while changing l_x has nearly no effects (37° phase shift of φ_y) on the spectra. The latter is highly desired indicating that the y-polarized response of our meta-atom is only sensitive to l_y but is very insensitive to l_x. Symmetry consideration implies that the same conclusion can be drawn if we interchange the indexes x and y.

In contrast, the polarization cross-talk effect is strongly enhanced if we remove the external wire loop from the resonator. As shown in Fig. 3.2e, in such a case the maximum variation of φ_y is only 161° due to changing l_y but can be as high as 83° due to changing l_x. Obviously, adding an external wire loop can significantly degrade the polarization cross-talk effects. The underlying physics of such an intriguing phenomenon can be understood by checking the field distributions on the cross-bar layers for the resonant modes associated with the $\vec{E}||\hat{y}$ polarization in two different meta-atoms, as shown in Fig. 3.2g. Due to the screening effect of the surrounding wire loop, the excited EM field is more strongly localized to the vicinities of the cross. In particular, the loop-bar coupling generates electric currents on the loop which significantly counteract the currents induced on the x-oriented bar, thereby making the mode rather insensitive to the parameter l_x. In contrast, non-negligible currents always exist on the x-oriented bar in the cross-only meta-atom even for the mode associated with the $\vec{E}||\hat{y}$ po-

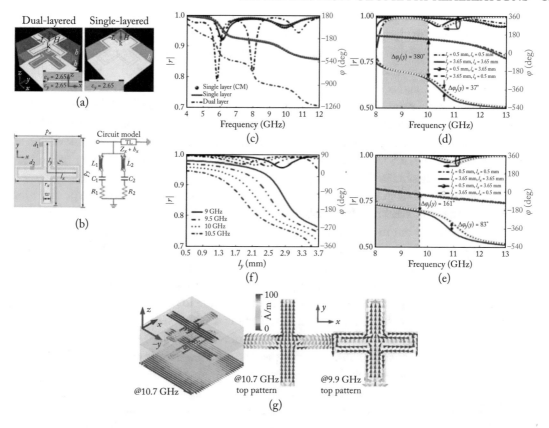

Figure 3.2: Illustration of the anisotropic meta-atoms without polarization cross-talk. (a) Topology of the dual-layer and single-layer anisotropic meta-atoms using composite crossbar and cross loop. (b) Illustration of geometrical parameters and equivalent circuit model. (c) Reflection coefficients of the single-layer and dual-layer meta-atoms in broad frequency spectrum, the circuit parameters are retrieved as $L_1 = 10$ nH, $C_1 = 0.05$ pF, $L_2 = 0.675$ nH, $C_2 = 0.108$ pF, $R_1 = 2.5\ \Omega$, $R_2 = 0.31\ \Omega$, $Z_c = 285.8\ \Omega$, and $h_o = 29.8°$. Reflection coefficients of the dual-layer meta-atom (d) with and (e) without external wire loop in cases of different l_x and l_y. (f) Reflection coefficients of the dual-layer meta-atoms as a function of l_y at different frequencies of 9, 9.5, 10, and 10.5 GHz when $l_x = 2$ mm. The residual geometrical parameters are $p_x = p_y = 8.3$ mm, $r_x = r_y = 8.1$ mm, $l_x = l_y = 3.65$ mm, $d_1 = d_2 = 0.25$ mm, and $w = 1$ mm. All results are calculated under the excitation of y-polarized incident waves. (g) Current distributions on metallic patterns of the cross-only (left and middle panel), and cross-loop dual-layer meta-atoms (right panel). In FDTD calculation of reflection magnitudes/phases of meta-atoms, we studied a unit cell containing a single meta-atom with periodic conditions applied at its four boundaries to mimic an infinite array, and a floquet port assigned at a distance 15 mm away from the x-y plane where the meta-atom is placed.

larization, generating the polarization cross-talk. Moreover, it is advisable to select a region far from the resonances to design our meta-atom with highly suppressed polarization cross-talk, as shown by the grey region of Fig. 3.2d. With the criterion to design our meta-atom known, we now show the bandwidth performance of the meta-atom. As illustrated in Fig. 3.2f, the reflection magnitude remains stable and is larger than 0.97 as l_y varies within 0.5–3.65 mm at 4 representative frequencies, indicating a satisfactory amplitude uniformity. More importantly, the four phase curves are almost parallel with each other and all exhibit good linear dependences on l_y in a large parameter region. One may easily expect that our meta-atom can work well in all frequencies lying in 9–10.5 GHz, where a full 360° phase coverage and near-unity magnitude can be simultaneously obtained by varying l_y.

Further calculations indicate that the decrease of bar-width (w) improves the polarization cross-talk property but weakens the capability of phase accumulation, see Fig. 3.3a, where the reflection coefficient of our meta-atom changes as a function of its and l_x. Again, the minimum and maximum values of l_x are set as $w/2$ and 3.65 mm, while l_y is fixed as 3.65 mm. Meanwhile, the resonant mode undergoes a slight red shift as w increases from 1 to 3 mm in the case of $l_x = 3.65$ mm. Therefore, increasing w deteriorates the irrelevance of l_x on $\xi_x(y)$ and $\xi_y(y)$ and that of l_y on $\xi_x(x)$ and $\xi_y(x)$, which is unfavorable for our design. However, further calculations indicate that the phase accumulation reduces when w is narrowed. For example, the phase difference as l_y changes from $w/2$ to 3.65 mm is 365° at 10 GHz for $w = 0.5$ mm, corresponding to 15° reduction relative to the case $w = 1$ mm. This is crucial for a large-array design since otherwise more layers should be added to guarantee the desired phase coverage and desirable polarization irrelevance, which are unfavorable to achieve low profile and high performance of devices. Balancing the two facts, we finally select a moderate value for $w (w = 1$ mm) in all designs. As shown in Fig. 3.3b, the phase is nearly a constant in the frequency regime 8–13 GHz when l_x increases from 0.5 to 3.65 mm, reinforcing that our meta-atom exhibits weak polarization cross-talking. The l_x-induced phase variation becomes non-negligible only at frequencies around 8.2 and 10.6 GHz where the two resonances are located and thus phase varies sharply. Above results indicate that it is advisable to select a region between the two highest resonances occurred when $l_x = 0.5$, $l_y = 3.65$ mm and when $l_x = l_y = 3.65$ mm, as shown by the grey region in Fig. 3.2d.

Before the formal design of the beam-control metadevices, we developed a general design methodology for any bifunctional devices under the framework of negligible polarization cross-talk. The design procedures mainly lie in four steps.

1. Design a subwavelength meta-atom with desirable polarization irrelevance and full 360° phase coverage within near-unity reflection amplitude and determine the basic structure parameters including periodicity p_x and p_y. The primary criterion as set previously is to decorrelate the variation of l_y with $\xi_x(x)$ and $\xi_y(x)$ and that of l_x with $\xi_x(y)$ and $\xi_y(y)$.

2. Determine the necessary aperture phase and amplitude distributions according to the target hybrid functionalities under two orthogonal polarizations. In this particular design,

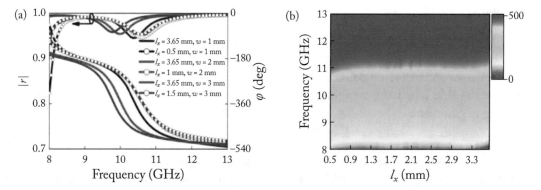

Figure 3.3: (a) Reflection coefficients of the anisotropic meta-atom as a function of frequency for different bar's width w. (b) Reflection phase of the anisotropic meta-atom as a function of frequency and l_x when $l_y = 3.65$ mm. The residual geometrical parameters are kept as $p_x = p_y = 8.3$, $r_x = r_y = 8.1$, and $d_1 = d_2 = 0.25$ mm.

three sets of phase gradients are involved, i.e., the linear phase gradient for beam steering, the parabolic phase gradient for focusing (single pencil beam) and the hyperbola one for multiple pencil beams. For linear or parabolic phase gradient, the 2D phase distribution along x or y direction can be theoretically synthesized. For the hyperbola gradient, it cannot be directly obtained through analytical equations but need cautious optimization through alternating projection method (APM) to achieve extremely low sidelobes.

The main procedure of quad-beam synthesis using APM is to search for the intersection between the set of possible radiation patterns (set A) of a metasurface and that of target patterns with idealized performance (set B), based on closed-loop iterative optimizations [147]. The tangential components of the EM field emitted from a metasurface is the sum of waves radiated from different meta-atoms,

$$A \equiv \left\{ T : T(u, v) = \sum_{(m,n) \in I}^{N} a_{m,n} e^{jk(P_{m,n}^x u + P_{m,n}^y v)} \right\}. \tag{3.3}$$

Here, I is the set of positions of all elements, $u = \sin \theta \cos \varphi$ and $v = \sin \theta \sin \varphi$ are the angular coordinates, $P_{m,n}^x$ and $P_{m,n}^y$ are positions of specific meta-atom along x and y direction, respectively, and $\alpha_{m,n}$ denotes the contribution from the meta-atom located at the position citation $\{m, n\}$, determined by both the excitation field and the response (reflection amplitude and phase) of the meta-atom itself. The target pattern requirements are specified by two masks, i.e., the multiple pencil beams with uniform amplitude and high gain; and low sidelobes with negligible radiations relative to the peak value. In the first mask, the -3 dB beamwidth and each main beam of target patterns are characterized by the lower- and upper-bound values ($M_L =$

0.707 and $M_U = 1$),

$$B \equiv \{T : T(u, v) = M_L(u, v) \le |T(u, v)| \le M_U(u, v)\}. \tag{3.4}$$

In the second mask, we define another upper bound M_U' at certain elevation angle θ. To minimize the side-lobe level, we require emitted fields in the side-lobe region must fulfill the following requirement:

$$B \equiv \{T : |T(u, v)| = M_U'\}. \tag{3.5}$$

The iterative optimization is considered to be converged and will be terminated when the cost function T_{adp} reaches a stable value. In this particular design, M_U' is restricted as an achievable value of -30 dB and the radiation pattern of the feed horn is modeled as $\cos^q(\theta)$ with $q = 8.6$,

$$T_{adp} = \sum_{u^2 + v^2 \le 1} \sum (|T(u, v)| - M_U')^2. \tag{3.6}$$

The synthesis consists of projecting the radiation patterns from set A to set B, and projecting the patterns back to the aperture magnitude and phase distribution (inverse Fourier Transform (FFT) algorithms). In the former case, the radiation patterns are rectified progressively until both sets are in good proximity, whereas in the latter case the phase and amplitude of elements across the aperture are dynamically renewed and finally reach the optimum distribution. In the reflective scheme, the quad-beam metasurface design is related only to phase synthesis since the element amplitude is determined by aperture size and illumination. Before the phase optimization, one should predetermine some initial parameters such as feed position F relative to metasurface, operation frequency f_0, aperture size D and element number $N = D/p$, elevation angles θ and azimuth angles φ that defining beam directions.

3. Choose an optimum strategy to cover the complete phase circle and obtain the phase-parameter database by FDTD parametric analyses. To guarantee sufficient precision, numerical interpolation is commonly adopted for available phase data with a mass of samples.

4. Determine the final metasurface by conducting a geometrical mapping process according to available phase distributions under two orthogonal polarizations based on a root-finding algorithm and phase-parameter database as shown in Fig. 3.2f. Thanks to the polarization irrelevance of the meta-atom, the geometrical parameters l_x and l_y can be separately determined by two polarization-dependent phase profiles $\varphi_x(x, y)$ and $\varphi_y(x, y)$, respectively.

Following above design procedures, we now employ the proposed anisotropic meta-atom to design the first versatile beam-control metadevice (denoted as sample I) to realize high-gain focused-beam and quad-beam emissions, trigged respectively, by x-polarized and y-polarized free-space excitations. As is shown in Figs. 3.4 and 3.5a, the metadevice is composed of a bi-functional metasurface I fed by a conical horn for excitation of both x and y-polarized incident

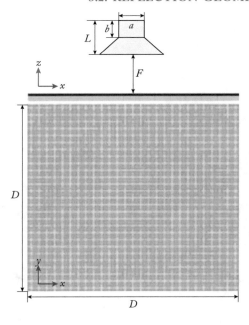

Figure 3.4: Topology of the emission system based on bifunctional metasurface I fed by a conical horn in both polarizations. The physical parameters of the horn are $a = 22.86$ mm, $b = 10.16$ mm, and $L = 30$ mm.

waves. The horn contains a standard x-band BJ-100 waveguide connecting to an open-end tapered waveguide. The aperture of the *tapered* waveguide was optimized as 44×24 mm^2 to have a gain of 10 dB with low sidelobes, and a good impedance match at $f_o = 10$ GHz. The square metasurface occupies an area of $D \times D = 224 \times 224$ mm^2. It is designed at $f_0 = 10$ GHz and composed of 27×27 anisotropic subwavelength meta-atoms each with a size of 8.3×8.3 mm^2. Moreover, a 2D distribution of varied l_x and l_y is clearly seen across the metasurface. For both excitations, the feed is placed $F = 125.5$ mm away from the center of the metasurface. The ratio of foci to diameter is $F/D = 0.56$, which is very beneficial to avoid the spillover radiation. To engineer the highly directive focused beam, a parabolic phase profile (Fig. 3.5c) is necessary which is determined by $\varphi(m,n) = \frac{2\pi(\sqrt{(mp)^2+(np)^2+F^2}-F)}{\lambda}$, where m and n label the position of a specific meta-atom along x and y axes, and F is the focal length. For easy characterization without loss of generality, the four pencil beams are chosen as $\varphi = 0°$, $90°$, $180°$, and $270°$, respectively, in both quadbeam syntheses and $\theta = 40°$ in this particular design.

Using the APM optimization process, we can synthesize the phase profile to exhibit desirable quad-beam emissions and low sidelobes. The final optimized radiation pattern and phase distribution are shown in Figs. 3.5b and 3.5d. Four pencil beams with uniform amplitude and high directivity can be clearly seen from the 3D radiation pattern of the metasurface. All beams

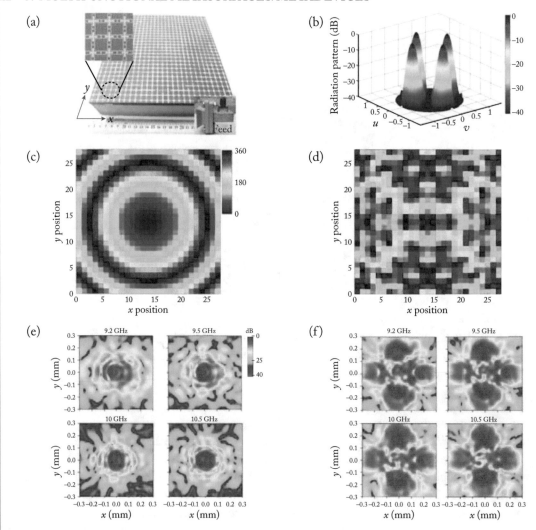

Figure 3.5: Design and near-field characterization of the bifunctional metasurface I with focused-beam and quad-beam directive emissions. (a) Photograph of the fabricated sample, the inset shows the conical feed horn and zoom-in view of the sample. (b) Theoretically calculated 3D quad-beam radiation patterns at 10 GHz. Objective phase distribution as a function of element position for (c) focused-beam and (d) quad-beam emissions under x-polarized and y-polarized excitations. Measured (e) E_x and (f) E_y distributions (real parts) of the bifunctional metasurface at 9.2, 9.5, 10, and 10.5 GHz in x-y plane under x-oriented and y-oriented polarization, respectively, with the probing waveguide, placed 90 mm away from the feed horn. The conical horn with an aperture size of 44×24 mm^2 was designed with 10 dB gain, low sidelobes, and good impedance match at $f_o = 10$ GHz.

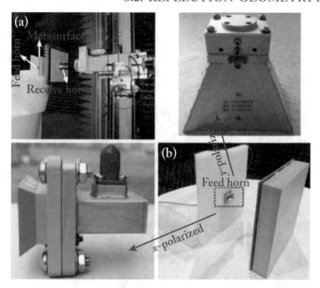

Figure 3.6: Illustration of the (a) near-field and (b) far-field experimental setup. In the far-field experiments, the feed horn and metasurface were fixed with a distance F on a big platform which can be rotated freely along its axial center. A receiver aligned with the metasurface is placed 9 m away to detect the radiated signal. In the near-field experiments, the local E_x and E_y field distributions (with both amplitude and phase) for two polarizations are recorded in x-y plane 70–90 mm away from the feed horn through a standard x-band waveguide which are fixed to an electronic step motor that moves with a step resolution of 5 mm by scanning a 2D area of 0.6×0.6 m² at the rear side of metasurface. Both near-field and far-field experiments are performed in an anechoic chamber.

are precisely directed to the angles at ($\varphi_1 = 0°$, $\theta = 40°$), ($\varphi_2 = 90°$, $\theta = 40°$), ($\varphi_3 = 180°$, $\theta = 40°$), and ($\varphi_4 = 270°$, $\theta = 40°$), which coincide well with our predetermined target. Moreover, the side-lobe level is below -38 dB, which is quite desirable. The phase distribution of the quad-beam array exhibits two-fold symmetry along x and y directions, yielding a 2D hyperbola-like phase profile with four regions separated by two diagonals. All dual-layer metadevices were fabricated with a standard printed-circuit-board (PCB) technology. The two dielectric boards were assembled together through adhesives, and then each metasurface was stick to a rigid foam to guarantee precise distance F between the feed and metasurface. The EM performance of the fabricated sample is characterized through both near-field and far-field measurements, utilizing experimental setup shown in Fig. 3.6.

Figures 3.5e and 3.5f depict the measured field distributions for E_x and E_y at four representative frequencies. As shown in Fig. 3.5e, the electric fields reflected from the metasurface are focused to a spot on the x-y plane at all frequencies studied. As such, the different optical paths

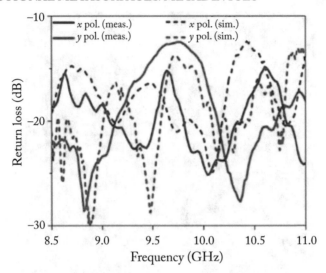

Figure 3.7: Simulated (dash line) and measured (solid line) return loss of the bifunctional meta-surface I under x and y polarization, respectively. A reasonable agreement of results is clearly observed between simulation and measurement in both cases. Slight deviations are partially attributable to the used adhesives and partially to the misalignment of feed and metasurface.

from the feed to metasurface are precisely collimated. The spot size is obviously decreased as frequency increases due to the shrinking of wavelength. On the other hand, four clear sub-spots with localized intensities are clearly observed in Fig. 3.5f, reinforcing the quad-beam emissions for the y-polarization observed from the far-field measurements (see Fig. 3.8). For both single-beam and quad-beam channels, the slightly distorted fields at the low- and upper-edge frequencies can be attributed to the worsened phases induced by the intrinsic Lorenz dispersions of resonant meta-atoms [148]. The functionalities of our device can be directly seen from Fig. 3.8a, where FDTD calculated 3D radiation patterns are shown at four representative frequencies. As is expected, we get completely different diagrams for wave-front control under two orthogonal polarizations (two channels). For x polarization (single-beam channel), highly directive pencil beam is clearly seen at all frequencies, where a peak gain of $G = 24.5$ dB is found at $f_0 = 10$ GHz and the aperture efficiency is calculated as 40.2% according to $\eta = \lambda^2 G/4\pi D^2$. However for y polarization (quadbeam channel), four symmetric pencil beams are clearly inspected at all frequencies studied, with a peak gain of near 17.5 dB at $f_0 = 9.7$ GHz for each beam. Therefore, an aperture efficiency of 33.9% is achieved for quadbeam channel according to $\eta = \lambda^2 \sum_{i=1}^{4} G_i/4\pi D^2$. The feed blockage poses negligible effect to device performance (see Fig. 3.7), where simulated and measured return loss is better than -12.3 dB across the entire observed band for two polarizations, indicating a decent impedance match. As shown in Fig. 3.8b, the gain for single-beam channel is better than 23 dB within 9.2–10.5 GHz. The aperture effi-

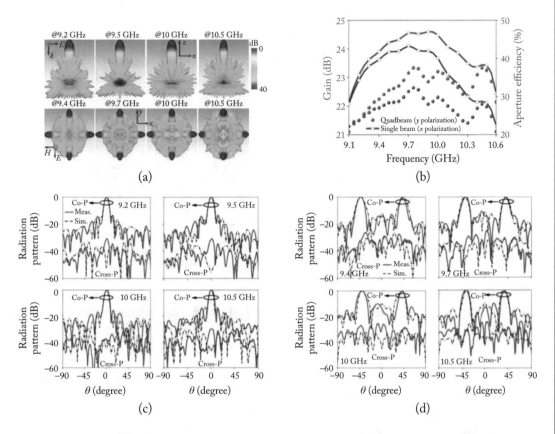

Figure 3.8: Far-field characterization of the emission system I based on bifunctional metasurface I under x-polarized and y-polarized normally incident plane waves. (a) Side view (tow row) and bottom view (bottom row) of the FDTD simulated far-field radiation patterns at four representative frequencies. (b) Measured gain and efficiency, the gain is a sum of four beams in quadbeam channel. (c) Simulated and measured radiation patterns in x-z plane at 9.2, 9.5, 10, and 10.5 GHz for x polarization. (d) Simulated and measured radiation patterns in x-z plane at 9.4, 9.7, 10, and 10.5 GHz for y polarization. Each pattern is normalized to the peak value of the main beam.

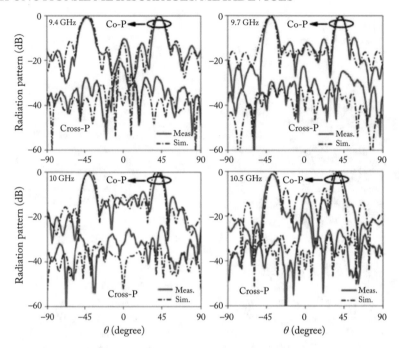

Figure 3.9: Simulated and measured radiation patterns in y-z plane of the emission system based on bifunctional metasurface I at 9.4, 9.7, 10, and 10.5 GHz, respectively, under y-oriented polarization.

ciency varies from 28.4–43.3% and the gain variation is less than ± 1 dB, leading to a 1 dB gain bandwidth of 13%. For quad-beam channel, the total gain is better than 22 dB while the gain variation is less than ± 1 dB within 9.4–10.6 GHz, where the aperture efficiency changes from 21–32.3%. The spillover loss and scanning loss of each oblique beam account for the relatively lower gain in quad-beam channel. A larger aperture size would be able to narrow the gain gap between these two cases.

Detailed performance of such bifunctional metasurface can be seen from Figs. 3.8c and 3.8d where the 2D radiation patterns of our device are shown in x-z (H) plane. Similar y-z (E) plane patterns can be found in Fig. 3.9. Among all cases, measurements are in reasonable agreement with simulations, further validating our design. For the y-polarization case, we observe two obvious pencil beams efficiently formed in each plane with almost equal intensity (tolerance less than ± 0.54 dB) and desirable symmetry. The half-power beam-width (HPBW) of a typical beam is calculated as $\sim 10°$, which is much less than that of the bare horn ($\sim 55°$). Moreover, the main beam is directed to 44°, 41°, 40° and 38° at 9.4, 9.7, 10, and 10.5 GHz, respectively. Such a frequency dependence of θ is quite physical since the metasurface becomes effectively enlarged in size though its physical size remains unchanged. The side-lobe level around boresight is better than -10.2 dB within 9.4–10.5 GHz and is around -20 dB for most elevation angles. In all cases, the cross-polarization is approximately 25 dB lower than the co-polarization peak. For x polarization, a highly directive pencil beam occurring at $\theta = 0°$ is clearly observed at all frequencies studied. The HPBW is about $7°$ and the measured cross-polarization is less than -29.8 dB in all cases. To sum up, the predicted two functionalities of our metasurface have been unambiguously demonstrated by both simulations and experiments.

We next design another bifunctional metadevice (denoted as sample II) using the same approach and the same type of meta-atoms. For this device, we integrate a linear phase profile and hyperbola-like phase profile into an ultrathin plate to realize two functionalities: beam-steering for x polarization and small-angle quadbeam emissions for y polarization. The steering angle is designed as $37°$ at 10 GHz by changing l_x while the directions of the quad-beams are engineered as $\theta = 30°$ at 10 GHz by changing l_y. Again, the twofold symmetric quadbeam phase distribution is synthesized by utilizing the APM algorithm in Matlab. The linear profile is determined by

$$\varphi(x) = \frac{2\pi(n-1)p_x \sin\theta}{\lambda},$$

where n is the element number belonging to $[1, N]$ and θ is elevation angle defining beam direction relative to the boresight.

The metadevice consists of 31×31 subwavelength meta-atoms and occupies an area of $D \times D = 257.3 \times 257.3$ mm^2, see the antenna topology shown Figs. 3.10 and 3.11a. l_x is periodically changed only along the x-direction, while a 2D change of l_y across the metasurface is clearly seen. For x polarization, a bigger tapered horn with an aperture size of 120×90 mm^2 is positioned at $F_x = 400$ mm away from the metasurface to guarantee an efficient plane-wave excitation. For y polarization, we used the previously utilized smaller horn to normally shine the metasurface at a distance of $F_y = 154.4$ mm $(F/D = 0.6)$. As expected in Fig. 3.11b, four beams with uniform intensity are achieved along x- and y-axis, respectively. All beams are precisely directed to $\theta = 30°$ off broadside and all sidelobes are suppressed below -32 dB, indicating an effective design and optimization process. As depicted in Fig. 3.11c, the designed phases are constant along y-direction while are progressively increased along x-direction with a constant gradient $(\varphi_{x(i+1)}(x) - \varphi_{x(i)}(x) = 60°, i = 1, 2, 3, 4, 5, 6)$. The metasurface using such a super-cell enables a beam steering behavior predicted by $\theta = \arcsin \lambda_p/6p_x$. As shown in Fig. 3.11d, a hyperbola-like phase profile across the aperture is clearly seen. As expected in Fig. 3.11e, our device works well within 9.6–10.7 GHz where the anomalous mode $(m = 1)$ dominants (the conversion efficiency is more than 72% and the peak efficiency is 95 and 89% in the numerical and experimental case at 9.7 GHz) while the normal mode $(m = 0)$ and first-order diffraction mode $(m = -1)$ are significantly suppressed, corresponding to a bandwidth of 11% relative to $f_0 = 10$ GHz. Such a band coincides well with that of metasurface exhibiting only l_x variation, further demonstrating the polarization irrelevance behavior. The operation bandwidth can be further identified as 9.6–10.7 GHz from Fig. 3.12b, where numerical and experimental reflection angles coincide well those predicted by the generalized Snell's law [21, 28, 30] and are

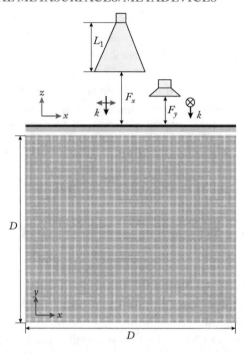

Figure 3.10: Topology of the emission system II based on bifunctional metasurface II fed by a bigger conical horn in x polarization while a smaller horn in y polarization. A bigger tapered horn positioned at $F_x = 400$ mm with an aperture size of 120×90 mm^2 and length of $L_1 = 170$ mm connected to a standard BJ-100 waveguide is utilized as the feed for excitation of x-polarized incident wave, whereas the previously utilized smaller conical horn normally shined the metasurface at a distance of $F_y = 154.4$ mm for excitation of y-polarized incident wave.

observed from 45.1–33.2° as frequency varies from 8.5–11 GHz. As portrayed in Fig. 3.11f, four localized spots with almost uniform E_y field intensities are symmetrically situated on the two axes at four representative frequencies, accounting for the efficiently synthesized quad pencil beams in the far-field region (Fig. 3.13). Out of the band 9.2–10.6 GHz, these spots become vague in the measured patterns.

Again, the effect of feed blockage to device performance can be negligible since the return loss in both cases is better than -11 dB across the entire band (see Fig. 3.12a), posing little penalty on antenna gain. The deteriorative matching at low ($f < 9.5$ GHz) and upper-($f > 10.5$ GHz) edge frequencies for x polarization is attributable to the increased specular reflections captured by the feed. The distorted linear phase gradient stemming from the inherent dispersion of resonant meta-atoms gives rise to the enhanced specular reflections at off working frequencies. However, the matching performance is not worsened at edge frequencies for y polarization. This is because little power can be captured by the small-sized horn due to large

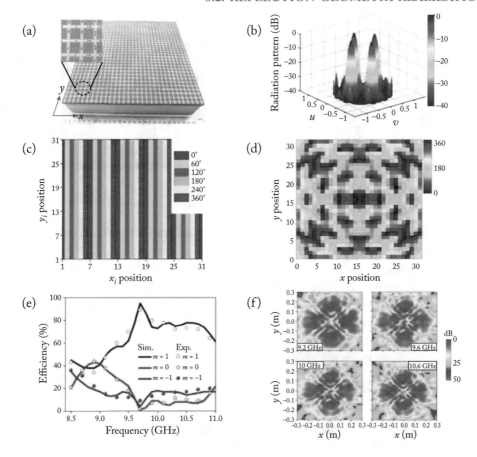

Figure 3.11: Design and near-field characterization of the bifunctional metasurface II with beam-steering and quad-beam directive emissions. (a) Photograph of the fabricated sample, the inset shows the zoom-in view of the sample and the sequentially changed l_x for the linear gradient is 3.65, 3.2, 2.6, 2.31, 2.1, and 1.83 mm. (b) Theoretically calculated 3D quad-beam patterns at 10 GHz. Objective phase distribution as a function of element position for (c) beam-steering and (d) quad-beam emissions under x-polarized and y-polarized excitations. (e) The diffraction efficiency of different orders as a function of frequency. The efficiency was defined as the ratio between reflected power ($\int_{-90°}^{\theta_{-1/2}} P(\theta)d\theta$, $\int_{\theta_{-1/2}}^{\theta_{1/2}} P(\theta)d\theta$, and $\int_{(\theta_1)/2}^{90°} P(\theta)d\theta$ with θ_{-1} and θ_1 being the reflection angles of $m=-1$ and $m=1$ mode) and the totally reflected power ($\int_{-90°}^{90°} P(\theta)d\theta$). (f) Measured E_y distributions (real parts) of the bifunctional metasurface in x-y plane at 9.2, 9.6, 10, and 10.6 GHz under y polarization, with the probing waveguide, placed 70 mm away from the feed horn. In the calculation of the scattering pattern in (e), to save the computational time we adopted a simplified system containing only one unit cell along the y-direction (with periodic boundary conditions set at its two boundaries) and all meta-atoms along the x-direction with open boundaries.

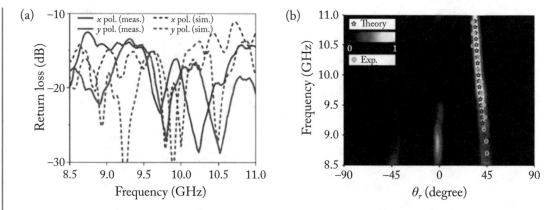

Figure 3.12: (a) Simulated (dash line) and measured (solid line) return loss of the bifunctional metasurface II within 8.5–11 GHz under x- and y-polarized excitation, respectively. Again, the reasonable agreement between simulations and measurements is observed in both cases. The observed discrepancies between simulation and experiments are due to the same reasons as discussed in Fig. 3.7. (b) The numerically calculated 2D contour of scattered power intensity $P(\theta r, \lambda)$ as a function of reflection angles $\theta_r (-90° < \theta_r < 90°)$ and frequency. Here, $P(\theta r, \lambda)$ is normalized to the peak intensity P_0 of a PEC ground with the same size, theoretical and experimental $\theta_r \sim f$ relations are afforded for comparisons.

angles of squint beams. As expected in Fig. 3.13a, a pure anomalous reflection beam is obtained at all frequencies studied, with both normal and first-order diffraction beams almost completely suppressed. Consequently, a high-gain scanning beam is achieved by placing a feed horn in front of the metasurface. Four pencil beams with symmetric patterns and uniform intensities are observed at a smaller elevation angle. As shown in Fig. 3.13b, a peak gain (efficiency) of 23 dB (21.6%) is achieved at 10 GHz for x polarization. A variation of gain of less than ±1 dB is observed from 9.5–10.5 GHz, corresponding to a bandwidth of 10%. The gain dips around 9.5, 9.7, and 10.2 GHz are because of the undesired reflection modes induced by the distorted phase profile. These modes tend to direct more energy into sidelobes. Nevertheless, the gain variation is within an acceptable level. For y polarization, the measured gain (aperture efficiency) varies from 23 dB (26.1%) at 9.1 GHz to 23.9 dB (23.7%) at 10.6 GHz with a peak value of 25.3 dB (38.2%) at 9.8 GHz. The relatively lower gain and efficiency in the beam-steering channel than that in quadbeam channel are likely caused by the spillover loss due to the finite dimensions of the sample. This is because in the former case we placed a bigger horn much farther from the metasurface to guarantee a plane-wave excitation. More power leaked into free space and thus the power captured by the metasurface was reduced. This issue can be substantially addressed by adding more supercells along with the aperture which is currently evaluated as 8.58λ at 10 GHz.

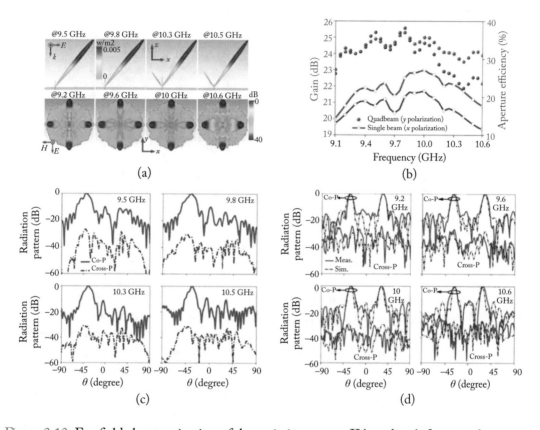

Figure 3.13: Far-field characterization of the emission system II based on bifunctional metasurface II under x-polarized and y-polarized normally incident plane waves. (a) FDTD-simulated far-field radiation patterns at four representative frequencies for x polarization (tow row) and y polarization (bottom row). (b) Measured gain and efficiency, the gain is a sum of four beams in quadbeam channel. (c) Measured radiation patterns in x-z plane at 9.5, 9.8, 10.3, and 10.5 GHz under x polarization, here numerical patterns in the beam-steering channel are not given due to computational memory constraints. (d) Simulated and measured radiation patterns in x-z plane at 9.2, 9.6, 10, and 10.6 GHz under y polarization. Each pattern is normalized to the peak value of the main beam.

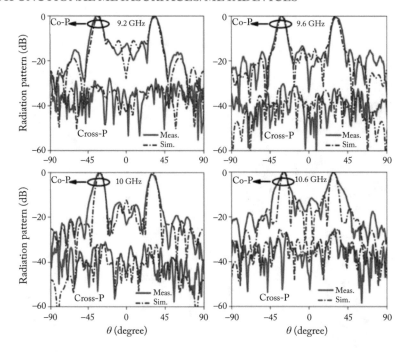

Figure 3.14: Simulated and measured radiation patterns in y-z plane of the bifunctional metasurface II at four representative frequencies of 9.2, 9.6, 10, and 10.6 GHz, respectively, under y-oriented polarization. As expected, numerical and experimental results coincide well and both illustrate two uniform narrowed beams with almost equal intensity (tolerance less than ± 0.4 dB).

The detailed performance of the emission system is further illustrated in Figs. 3.13c and 3.13d, where the radiation patterns are plotted in x-z plane in two different polarizations. Similar patterns can be observed in y-z plane, see Fig. 3.14. From Fig. 3.13c, we found that the beam-steering angle changes from -39–$-35°$ as frequency varies from 9.5–10.5 GHz. The averaged HPBW is around 13° and the sidelobes are better than -10 dB at most frequencies and the cross-polarization is better than -26.7 dB. As shown in Fig. 3.13d, satisfactory agreement is observed between simulations and measurements at all frequencies studied. Two uniform narrowed beams with almost equal intensity are clearly seen (tolerance less than ± 0.25 dB). The patterns do not change appreciably from 9.2–10.6 GHz, indicating a stable and robust multibeam emission. All measured cross-polarization is 24.5 dB below the peak intensity. The HPBW is about 8° which is narrower than our first device, thanks to a larger aperture size of the present sample.

3.3 TRANSMISSION-GEOMETRY REALIZATIONS: FOCUSING AND BEAM BENDING

In this section, we design a bifunctional metadevice working in transmission geometry to realize distinct functionalities with very high efficiencies. First, we discuss our approach to design such a transmissive device realizing a focusing lens and a beam deflector for incident waves with $\vec{E}\|\hat{y}$ and $\vec{E}\|\hat{x}$ polarizations, respectively. For a system with global mirror symmetry, the $\hat{x}$ (or $\hat{y}$)-polarized incident wave can only respond to the phase distribution $\phi_{xx}(x, y)$ (or $\phi_{yy}(x, y)$), thus we can govern $\phi_{xx}(x, y)$ and $\phi_{yy}(x, y)$ to realize two different functionalities. To achieve focusing for $\hat{y}$ polarization, $\phi_{yy}(x, y)$ should satisfy a parabolic distribution

$$\phi_{yy}(x, y) = k_0 \left(\sqrt{F^2 + x^2 + y^2} - F \right) \tag{3.7}$$

with F denoting the focal length and $k_0 = \omega/c$. To realize beam bending for the $\hat{x}$ polarization, $\phi_{xx}(x, y)$ must satisfy a linear distribution,

$$\varphi_{xx}(x, y) = C_1 + \xi \cdot x, \tag{3.8}$$

where C_1 is a constant and ξ is a predesigned phase gradient that can determine the bending angle via the generalized Snell's law [21, 28, 30].

Second, we discuss how to design a transmissive meta-atoms with varied transmission phases (ϕ_{xx} and ϕ_{yy}) and nearly unit transmission amplitudes. Completely different from the reflective meta-atoms, the design of transmissive meta-atoms faces more challenges since the transmission phase is usually linked with transmission amplitude. And both a full-range of phase modulation over 2π and a high transmission are required to guarantee a free wavefront manipulation. We know that a resonant element composing of a metallic mesh coupled with a metal patch (see inset to Fig. 3.15a) can support a perfect EM transmission at a typical frequency, due to the interaction between the Lorentz resonator (i.e., the metal patch) and the opaque background (the metallic mesh) [149]. However, the transmissive property, such as the transparency window and the associated transmissive phase variation range, is quite limited. We solve these problems by using a cascaded multiplayer system. For four-layer composite meta-atom as shown in Fig. 3.15b, both the transparency window and the transmission-phase variation range are significantly enlarged resulting from the mutual interactions among these cascaded particles. In our meta-atom, the presence of the metallic mesh can significantly reduce the mutual couplings between adjacent meta-atoms, making our design robust and reliable [150]. As depicted in Fig. 3.15b, within a wide transparency frequency window of 9.1–11.7 GHz, the transmission phase ϕ undergoes a nearly 360° range while exhibiting a high transmission amplitude of $|t| > 0.8$. Figures 3.16a,b depict how the transmission amplitude $|t_{xx}|$ and phase ϕ_{xx} of the meta-atom depend on the parameters a and b, at the working frequency $f_0 = 10.5$ GHz. We note that $|t_{xx}|$ always keeps at a high value (> 0.8) as we vary the two parameters (see Fig. 3.16a), thanks to our special design of the multilayer meta-atom, but the phase ϕ_{xx} is quite sensitive to the variation of parameter a. Again, we note that the dependences of $|t_{yy}|$ and phase ϕ_{yy} on a and b can

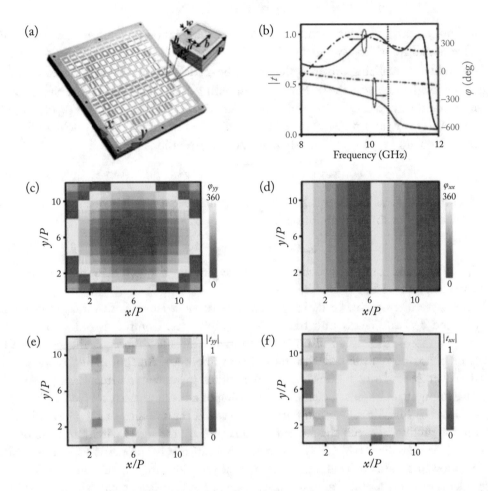

Figure 3.15: Design of the bifunctional metasurface working in transmission geometry. (a) Picture of the fabricated sample. Inset shows the typical meta-atom composed of four metallic layers separated by three F4B spacers ($\varepsilon_r = 2.65 + 0.01i$, $d = 1.5$ mm). Other parameters are fixed as $P = 11$ mm, $w = 0.5$ mm, $h = 36$ μm. (b) FDTD simulated transmission amplitude (blue lines) and phase (red lines) for the metasurface composing of the periodic array of meta-atoms with $a = 5$ mm and $b = 8.8$ mm, shining by $\hat{y}$-polarized (solid lines) and $\hat{x}$-polarized (dot lines) incident waves, respectively. (c–f) FDTD simulated ϕ_{yy}, ϕ_{xx}, $|t_{yy}|$, and $|t_{xx}|$ distributions at each meta-atom of the designed/fabricated transmissive metadevice at the frequency $f_0 = 10.5$ GHz.

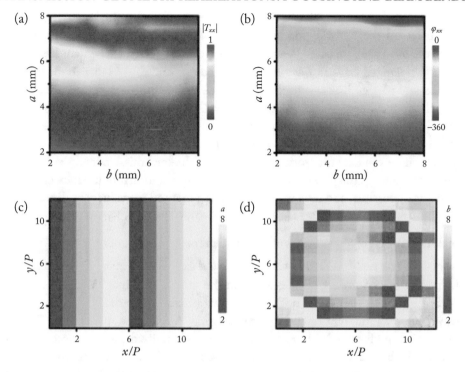

Figure 3.16: The maps of transmission amplitude and transmission phase of the meta-atom and detailed structural parameters of the transmissive metadevice. (a) Transmission amplitude $|t_{xx}|$ map and (b) transmission phase ϕ_{xx} as functions of a and b. The other parameters of the transmissive meta-atom are fixed as $P = 11$ mm, $w = 0.5$ mm, $h = 36$ μm, $d = 1.5$ mm, and $e_r = 2.65 + 0.01i$. (c) a and (d) b distributions for the designed/fabricated transmissive meta-surface. The parameter units of a and b are in mm.

be retrieved from Figs. 3.16a,b based on symmetry considerations. With these information in hand, we follow the general design strategy to determine the geometric parameters (a and b) of all meta-atoms involved in our transmissive metadevice with $\xi = 0.43k_0$ and $F = 60$ mm. Figures 3.16c,d depict the distributions of $a(x, y)$ and $b(x, y)$ in the finally designed metadevice, the corresponding transmission phases (ϕ_{xx} and ϕ_{yy}) and transmission amplitudes ($|t_{yy}|$ and $|t_{xx}|$) at each meta-atom are plotted in Figs. 3.15c–f, respectively. Based on the design, we fabricated a sample of bifunctional metasurface with its picture shown in Fig. 3.15a. The sample consists of 12×12 meta-atoms with a total size of 132×132 mm^2.

Third, we experimentally examine its functionalities with two steps. In the first step, we investigate its focusing functionality. Shining normally onto our sample by a wide band horn antenna, we use a monopole antenna (~ 30 mm) to detect the electric-field distributions at the transmission part of the sample through an automatically controlled scanning mapper system [30]. To record the transmission amplitude and phase information, both the monopole antenna and the horn are connected to a vector-field network analyzer (Agilent E8362C PNA). Shining by a $\hat{y}$-polarized wave at $f_0 = 10.5$ GHz, the measured $Re[\vec{E}]$ distributions at both xoz and yoz planes are plotted in Fig. 3.17a. We can see that the incident waves are converged to a focal point. The focal length is calculated as $F = 64$ mm in measurement, which is in good agreement with the numerical simulation result ($F = 62$ mm), as shown in Fig. 3.20. Here, the deviation from the theoretical value ($F = 60$ mm) is mainly caused by the finite-size effect of our sample. Then we evaluate the working efficiency of the transmissive meta-lens, which is defined as the ratio between the power carried by the focal spot and that of the incident one. Direct measurement is impossible, and here we use a two-step method to evaluate it. The efficiency can be calculated by $\eta_{foc} = \frac{P_{tran}}{P_{tot}} \times \frac{P_{foc}}{P_{tran}}$, where $\frac{P_{tran}}{P_{tot}}$ represents the ratio between the power carried by the total transmitted waves and that of the incident one and $\frac{P_{foc}}{P_{tran}}$, defined by $\frac{P_{foc}}{P_{tran}} = \frac{\oint \hat{n} \cdot \vec{S} \, ds_1}{\oint \hat{n} \cdot \vec{S} \, ds_2}$, can be evaluated rigorously in FDTD simulations and can approximately be obtained in experiments by replacing $\hat{n} \cdot \vec{S}$ with $|\vec{E}|^2$ measured on the focal plan. First, We evaluate the former term $\frac{P_{tran}}{P_{tot}}$. Since the meta-lens does not exhibit any translation-invariance symmetry, the transmitted waves go to all different directions after passing through the focal point. Therefore, we should in principle perform a 3D integration over the transmitted waves to accurately evaluate the efficiency, which is very difficult to perform experimentally. Considering the 2D symmetry in our focusing geometry, here we chose two typical planes to perform 2D integrations over the transmitted waves, which are much easier to do in our experiments. Shining the transmissive metasurface with a $\hat{y}$-polarized incident plane wave at $f_0 = 10.5$ GHz, we measured the power distributions of the scattered waves (at both transmission and reflection sides) on both yoz and xoz planes, and depicted the patterns in Figs. 3.18a,c, respectively. The corresponding FDTD-simulated patterns are depicted in Figs. 3.18b,d for comparisons. The discrepancies between measurement and FDTD results originate from the imperfect wave-front profile of the incident wave emitted from the horn antenna and the imperfection of the fabricated sample. To estimate the reference power (i.e., P_{tot}), we simulated and measured the scattered-field distributions of a metallic slab with the same size as our metasurface (see Figs. 3.18e,f). For each pattern depicted in Figs. 3.18a–d, we integrated the total power at the transmission side and then computed the ratio between that integrated value and the corresponding reference values obtained by integrating the power distributions shown in Figs. 3.18e,f. The obtained ratios are depicted in each figure. We note that these values are quite close to each other, indicating that the asymmetry caused by the square shape of the meta-lens is not significant. In addition, the FDTD simulated values are also quite close to the measured ones. We also estimated the dielectric losses in different cases, which are found about 1% in each case, indicating that the dielectric losses are quite small in our case. Second, we evaluated the latter term $\frac{P_{foc}}{P_{tran}}$ by integrating the measured $|\vec{E}|^2$ inside the focal-spot and the entire plane, respectively. Shining the sample with

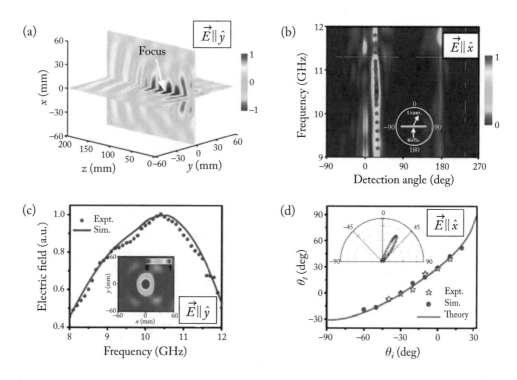

Figure 3.17: Characterizations of the transmissive bifunctional metasurface. (a) Measured $Re[\vec{E}]$ distributions on both *xoz* and *yoz* planes as our metasurface is shined by a $\hat{y}$-polarized microwave at $f_0 = 10.5$ GHz. (b) Measured and simulated results of $\vec{E}$-field amplitude at the focal points of different frequencies. Inset shows the measured $\vec{E}$-field intension at $z = 64$ mm plane of $f_0 = 10.5$ GHz. All the $\vec{E}$-field amplitudes are normalized to the corresponding maximum value obtained in the spectrum. (c) Measured angular power distribution map as functions of frequency and detection angle as our sample is shined by $\hat{x}$-polarized incident microwaves normally. (d) Measured (blue open star symbols), FDTD simulations (red solid circle symbols) and theory (green lines) anomalous-refraction angle θ_t against the incident angle θ_i as our sample shined by an $\hat{x}$-polarized plane wave at $f_0 = 10.5$ GHz. Inset depicts the measured and FDTD simulated angular power distributions of the transmitted waves as our sample illuminated by an $\hat{x}$-polarized wave at $f_0 = 10.5$ GHz.

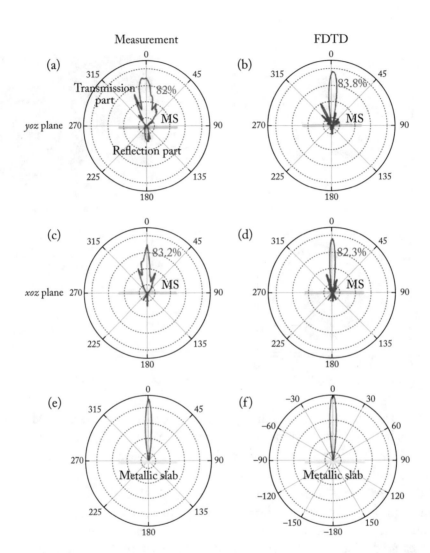

Figure 3.18: (a, c) Measured and (b, d) simulated power distributions of the scattered waves on the *yoz* and *xoz* planes, respectively, for our transmissive metasurface illuminated by a $\hat{y}$-polarized incident plane wave at the frequency $f_0 = 10.5$ GHz. (e) Measured and (f) simulated power distributions of the scattered waves for a metallic slab with the same size of the metasurface illuminated by a $\hat{y}$-polarized incident plane wave at the frequency $f_0 = 10.5$ GHz.

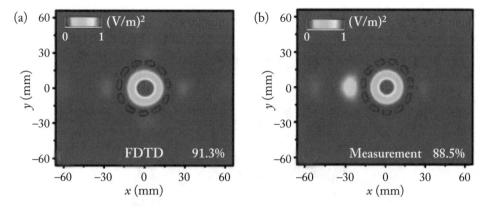

Figure 3.19: (a) FDTD simulated and (b) measured $|\vec{E}|^2$ distributions on $z = 62$ mm and $z = 64$ mm, for our transmissive metasurface illuminated by a $\hat{y}$-polarized incident plane wave at the frequency, respectively.

a $\hat{y}$-polarized incident plane wave at $f_0 = 10.5$ GHz, we measured the $|\vec{E}|^2$ distribution at the focal plane $z = 64$ mm and depicted the results shown in Fig. 3.19. The area of focal spot (red dash line in Fig. 3.19) is chosen within the region defined by the first zero. Calculations based on measured $|\vec{E}|^2$ distributions yield that $\frac{P_{foc}}{P_{tran}} \approx 88.5\%$, which is close to 91.3% estimated based on FDTD simulations. From the results obtained by the two-step analyses, we can obtain the working efficiency of our meta-lens, which is in the range of 72.5–73.6% based on experimental measurements and lies in the range of 75.1–76.5% based on FDTD simulations. As shown in Fig. 3.17b, the working bandwidth of our lens is about 2.9 GHz (8.7–11.6 GHz), evaluated by the FWHM of the intensity of $\vec{E}$-field. Within this frequency bandwidth, the focal-spot size changes in the range of 17–20.5 mm while the focal length varies in the range of 43–70 mm, as shown in Fig. 3.21.

In the second step, we characterize the beam-deflection functionality of our metadevice. We measured the angular power distributions at both transmission and reflection sides of our metasurface as illuminating by an $\hat{x}$-polarized microwave, as the schematics are shown in the inset to Fig. 3.17c. We note clearly that within a wide frequency interval (10.2–11.3 GHz), most transmitted waves are deflected to an anomalous angle, which agrees well with the generalized Snell's law $\theta_t = \sin^{-1}(\xi/k_0)$ (solid stars in Fig. 3.17c). At the target working frequency $f_0 = 10.5$ GHz, the normal-mode transmission is completely depressed and the anomalous transmission reaches a maximum. The absolutely working efficiency, defined as the ratio between the power carried by the anomalously deflected beam and that of the incident beam, is measured as 72% and simulated as 73%. The missing power is carried partly by the reflection mode, partly by the dielectric loss and partly by the other radiation modes. Figure 3.17d illustrates the $\theta_t \sim \theta_i$ the relation for the anomalous refraction at the frequency $f_0 = 10.5$ GHz,

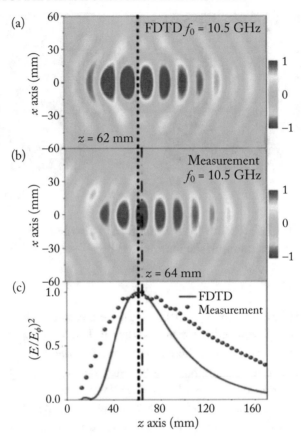

Figure 3.20: (a) Simulated and (b) measured $Re(\vec{E})$ distributions at the *xoz* plane for the transmissive metasurface illuminated by a $\hat{y}$-polarized wave at $f_0 = 10.5$ GHz. (c) Simulated (blue solid circles) and measured (red line) energy distribution along z-axis for transmissive meta-lens.

with θ_t and θ_i being the refraction and incident angles, respectively. We can see clearly that the measured results (blue star), the FDTD simulations (solid red circle) and the theoretical analysis based on the generalized Snell's law (green line) are in perfect agreement.

3.4   FULL-SPACE WAVE-CONTROL MULTIFUNCTIONAL METASURFACES

We discuss the design of bifunctional metasurfaces in transmission and reflection geometries, respectively, in the last two sections. However, the wave-manipulation capabilities of these metasurfaces are far less explored, since they work either in pure reflection mode (Fig. 3.22a) [13, 34, 54, 151] or pure transmission mode (Fig. 3.22b) [26, 38, 152, 153], leaving the other half-space

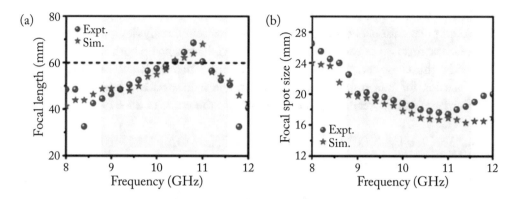

Figure 3.21: Quality of the focal spot for the designed transmissive meta-lens. (a) Simulated and measured focal length and (b) focal-spot size for the transmissive meta-lens for our metasurfaces illuminated by $\hat{y}$-polarized plane waves.

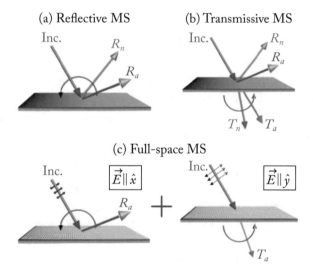

Figure 3.22: Working mechanism and advantages of the full-space metasurface. Conventional metasurfaces working in (a) reflection or (b) transmission geometries, which is seriously limited to the restricted working space, low efficiency due to multi-mode generations and locked phase gradients for the transmitted and reflected waves. (c) Our proposed full-space metasurface can control EM wavefronts at both sides of the metadevice, determined by the incident waves with different polarizations.

uncontrolled. In this section, we propose a novel strategy to design metasurfaces that can manipulate the EM wave-fronts in full space with very high efficiencies (see Figs. 3.22c,d), dictated by the incident polarizations. For demonstration, we experimentally designed three microwave metasurfaces. The first two can achieve beam bending and focusing on both sides of the metasurfaces, while the third case can combine these two bifunctionalities (wave-bending for reflected wave and focusing for transmitted wave) on one single metasurface. More importantly, all of those metasurfaces exhibit very high efficiencies within the range of 85–91%.

3.4.1   CONCEPT AND META-ATOM DESIGN OF FULL-SPACE METASURFACE

We start by describing our mechanism to achieve the full-space wavefront manipulation and design an appropriate meta-atom. Here, we consider a system with mirror symmetry for simplification. As we know, the EM property of such a system can be described by two diagonal Jones matrices

$$R = \begin{pmatrix} r_{xx} & 0 \\ 0 & r_{yy} \end{pmatrix}$$

and

$$T = \begin{pmatrix} t_{xx} & 0 \\ 0 & t_{yy} \end{pmatrix},$$

with r_{xx}, r_{yy}, t_{xx}, and t_{yy} denoting the reflection/transmission coefficients along two principal axes $\hat{x}$ and $\hat{y}$, respectively. In a lossless system, we can obtain $|r_{xx}|^2 + |t_{xx}|^2 = 1$ and $|r_{yy}|^2 + |t_{yy}|^2 = 1$ due to energy conservation. In order to realize independent yet highly efficient manipulation on both the transmitted and reflected waves, we need the meta-atom to be completely transmissive and reflective for both polarizations, respectively (i.e., $|t_{xx}| = 0$, $|r_{xx}| = 1$, and $|t_{yy}| = 1$, $|r_{yy}| = 0$). Moreover, the corresponding phases ϕ^r_{xx} and ϕ^t_{yy} should be freely tuned by changing the structural parameters of the meta-atoms. With these meta-atoms, we can construct metasurfaces exhibiting required phase distributions (i.e., $\phi^r_{xx}(x, y)$ and $\phi^t_{yy}(x, y)$) to realize different functionalities by controlling reflected and transmitted wave-fronts, under $\hat{x}$- and $\hat{y}$-polarized excitations, respectively.

Then we design a realistic meta-atom that exhibits the mentioned full-space wavefront manipulation characteristics. As shown in Fig. 3.23a, the designed meta-atom consists of four cross metallic layers which are separated by three 1.5-mm-thick F4B spacers ($0.16\lambda_0$). At x-direction, the first two layers have finite yet tuned lengths while the metallic stripes are continuous at the bottom two layers. The continuous metallic stripes can couple with the upper two metallic resonators to create magnetic resonances at frequencies dictated by the geometrical parameters at x-direction, which leads to the strong modulation on the reflection phases ϕ_{xx} to undergo a continuous variation between $-180°$ to $180°$ when passing through the magnetic resonance. However, at y-direction, the identical four-metallic-layer structures, providing mutual couplings among these resonators, are used to improve the transmission amplitude $|t_{yy}|$ to nearly 1 and enlarge the variation range of ϕ_{yy} to cover 360°, while being very insensitive to r_{xx}, since the normally incident waves with $\hat{y}$ polarization can only "see" the phase along the y-direction. Now the advantages of our design are quite clear. Those x-orientated continuous metallic stripes at the bottom two layers can essentially work as an effective optical grating to efficiently reflect the $\hat{x}$-polarized wave and transmit only the $\hat{y}$-polarized wave at the designed frequency. We next fabricated a sample with a dimension of 330 mm × 330 mm (see Fig. 3.23b for its pictures) consisting of a periodic array of meta-atoms. In this case, only one magnetic resonance generates since the metallic stripe in the second layer is also continuous ($d_2 = 11$ mm). Two magnetic resonances can appear if the metallic stripe in the second layer exhibits a finite length. Therefore, we can obtain expanded freedom to design our meta-atom by tuning both d_1 and d_2. The transmission and reflection properties of the sample are characterized by microwave experiments. Figures 3.23c,d illustrate the good agreements between the measured and simulated results, in both terms of amplitude and phase profile. As expected, within a frequency interval (7–13 GHz), a totally reflective band ($|r_{xx}| > 0.92$) can be observed with the reflection phase variation range larger than 360° for the $\hat{x}$-polarization. Meanwhile, a high transmission ($|t_{yy}| > 0.84$) with ϕ_{yy} covering a nearly 360° variation range is obtained within a wide transparency window ($\sim$ 7.5–13 GHz) for the $\hat{y}$-polarization.

Figure 3.23 suggests that our meta-atom is an ideal particle to design metasurfaces achieving the full-space wave-front control. To verify our design, we present in Fig. 3.24 the detailed EM response maps of the meta-atom (e.g., ϕ_{xx}^r, $|r_{xx}|$, ϕ_{yy}^t, and $|t_{yy}|$), based on which we can easily fix the structural details of meta-atoms at different positions in a metasurface according to its desired phase distributions. Figures 3.24a–d illustrate, respectively, how ϕ_{xx}^r, $|r_{xx}|$, ϕ_{yy}^t, and $|t_{yy}|$ vary against the parameters a, d_1, and d_2, with the frequency fixed at $f_0 = 10.6$ GHz. Obviously, ϕ_{xx}^r is sensitive to d_1 and d_2 but insensitive to a, while ϕ_{yy}^t behaves just oppositely. These nearly independent control abilities make our realistic meta-atom design relatively easy. By changing the structural parameters appropriately, the variations of ϕ_{xx}^r and ϕ_{yy}^t can cover the whole 360° range, while simultaneously $|r_{xx}|$ and $|t_{yy}|$ remain at very high values ($|r_{xx}| > 0.92$, $|t_{yy}| > 0.84$), which guarantees the high working efficiency of the designed metasurface.

3.4.2   FULL-SPACE BEAM DEFLECTOR

We first design a full-space and high-efficiency beam deflector with the proposed meta-atom. To achieve this goal, ϕ_{xx} and ϕ_{yy} should satisfy the following linear distributions:

$$\begin{cases} \phi_{xx} = C_0 + \xi_1 \cdot x \\ \phi_{yy} = C_1 + \xi_2 \cdot x \end{cases} \tag{3.9}$$

with C_0 and C_1 being phase constants, ξ_1 and ξ_2 being the reflection and transmission phase gradients provided by the metasurface, respectively, which determine the bending angles of the anomalous reflection and transmission beam based on the generalized Snell's law [21, 28, 30],

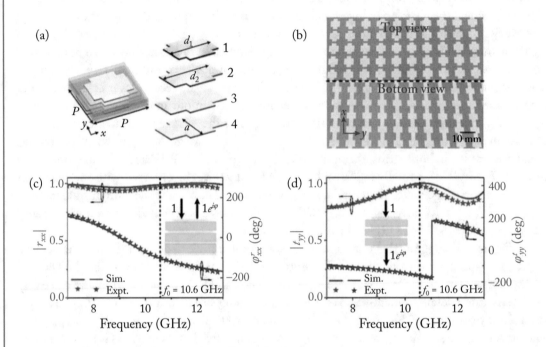

Figure 3.23: Design and characterization of the proposed meta-atom. (a) Schematic of the designed meta-atom. The typical meta-atom composed of four metallic layers separated by three F4B spacers ($\varepsilon_r = 2.65 + 0.01i$, $h = 1.5$ mm). Other parameters are $w_1 = 5$ mm, $w_2 = 4$ mm, $P = d_3 = 11$ mm. (b) Pictures of fabricated metasurface with top view and bottom view. Measured and FDTD simulated spectra of (c) reflection coefficient and (d) transmission coefficient for the metasurface constructed by a periodic array of meta-atoms with $a = 6.3$ mm, $b_1 = 9$ mm, and $b_2 = 11$ mm, under excitations of (c) $\hat{x}$-polarized and (d) $\hat{y}$-polarized incident waves, respectively. Inset depicts the schematics of working principles of the metasurface as (c) a reflective blockage and (d) a transparency window under excitations of (c) $\hat{x}$-polarized and (d) $\hat{y}$-polarized incident waves, respectively.

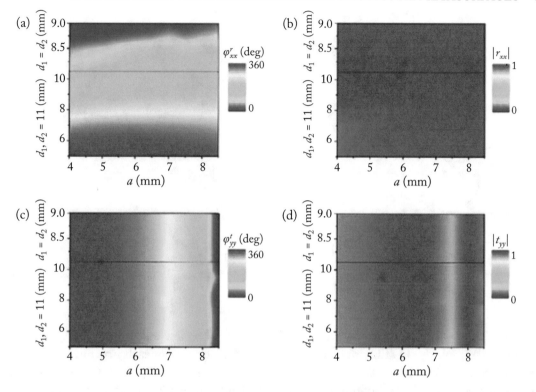

Figure 3.24: EM response map of the designed meta-atom. (a) Reflection phase ϕ_{xx}^r and (b) the amplitude $|r_{xx}|$ maps of the meta-atom as functions of a and d_1 (d_2) for $\hat{x}$-polarized incidence. (c) Transmission phase ϕ_{yy}^t and (d) amplitude $|t_{yy}|$ maps of the meta-atom as functions of a and d_1 (d_2) for $\hat{y}$-polarized incidence. Here, the frequency is fixed as $f_0 = 10.6$ GHz.

respectively. We optimize six meta-atoms as a supercell based on the parameter maps shown in Fig. 3.24 to make their phases satisfying Eq. (3.9), and $\xi_1 = -0.43k_0$ and $\xi_2 = 0.43k_0$, with $k_0 = \omega/c$ being the propagation constant. Figures 3.25c and 3.25d depict the FDTD simulated distributions of reflection/transmission amplitude/phase of the designed metasurface. We note that the designed/fabricated ϕ_{xx} and ϕ_{yy} coincide well with the theoretical values, meanwhile these meta-atoms exhibit competitive reflection (> 0.93) and transmission amplitudes (> 0.86), indicating high working efficiencies of our metadevices. The fabricated metadevice contains 30×30 meta-atoms with a total size of 330×330 mm^2, as the pictures seen in Figs. 3.25a,b.

With the sample in the band, we experimentally examine its EM functionalities. First, we examine the anomalous beam-deflection functionality at the reflection part of our designed metasurface. Shining an $\hat{x}$-polarized incident wave normally onto our metasurface, we measure the normalized angular distributions of scattered-wave intensities with two horn antennas, and

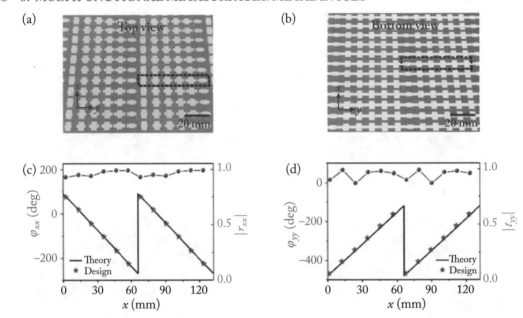

Figure 3.25: Design of a full-space beam deflector. (a) Top view and (b) bottom view of the fabricated sample. (c) FDTD simulated ϕ_{xx} and $|r_{xx}|$ distributions of each structural unit within two supercells, with black solid line representing $\phi_{xx} = C_0 + \xi_1 \cdot x$ ($\xi_1 = -0.43k_0$). (d) FDTD simulated ϕ_{yy} and $|t_{yy}|$ distributions of each structural unit within two supercells, with black solid line representing $\phi_{yy} = C_1 + \xi_2 \cdot x$ ($\phi_2 = 0.43k_0$).

plot the reflection and transmission maps in Figs. 3.26a,b, respectively. The signals are normalized to the ones received by replacing our metadevice with a metallic plate of the same size. Experimental results show clearly that, within a frequency interval ranging from 10.4–11 GHz, most of the $\hat{x}$-polarized incident waves are reflected to anomalous angles, which agrees well with the theoretical values predicted by the generalized Snell's law $\theta_t = \sin^{-1}(\xi/k_0)$ (solid stars in Fig. 3.26a) [21, 28, 30]. The best performance appear at our target frequency of $f_0 = 10.6$ GHz, where all undesired modes are almost totally suppressed, evidenced by the good agreements between FDTD simulations (Figs. 3.27a,b) and experimental results (Figs. 3.26a,b) on the far-field patterns. It's worth noting that the maximum absolute efficiency of our metadevice based on experimental results can reach as high as 91% at the frequency of $f_0 = 10.6$ GHz, which is very close to simulation value of 93%, as shown in Fig. 3.26c. Here, we evaluate the absolute efficiency by the ratio between the power over the angle regions spanned by the anomalous modes and the total incident power. The slight difference between the measured and simulated results comes from inevitable fabrication errors and imperfections of the incoming wave-fronts.

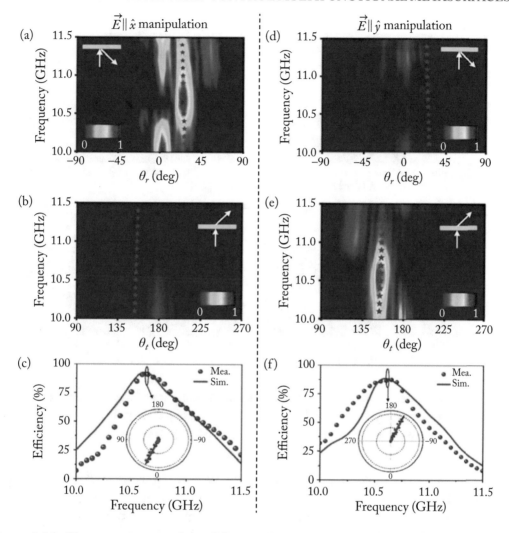

Figure 3.26: Characterizations of the full-space beam deflector under normal incidence. (a–c) Reflection manipulation and (d–f) transmission manipulation under illuminations of incident waves with polarizations of (a–c) $\vec{E}||\hat{x}$ and (d–f) $\vec{E}||\hat{y}$. Measured scattered-field intensities (color map) in (a, d) reflection and (b, e) transmission sides vs. frequency and the detecting angles θ_r/θ_t, for our metasurface illuminated by normally incident waves with (a, b) $\hat{x}$-polarization and (d, e) $\hat{y}$-polarization. FDTD simulated and measured absolute efficiencies of (c) anomalous reflection and (f) anomalous refraction for our metasurface under normal incidence with (c) $\vec{E}||\hat{x}$ and (f) $\vec{E}||\hat{y}$, respectively. Inset shows the measured and FDTD-simulated angular distributions of the waves (c) reflected and (f) transmitted by the metasurface, illuminated by (c) $\hat{x}$-polarized and (f) $\hat{y}$-polarized incident waves at $f_0 = 10.6$ GHz.

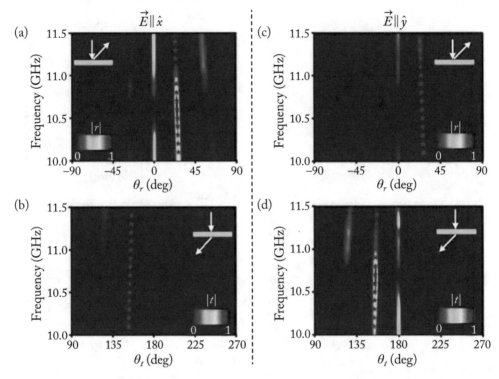

Figure 3.27: FDTD-simulated scattered-field intensities (color map) in (a, c) reflection and (b, d) transmission space vs. frequency and the detecting angles of our meta-beam-deflector shined by normal-incidence wave with (a, b) $\vec{E}||\hat{x}$ and (d, e) $\vec{E}||\hat{y}$ polarizations, respectively.

Second, we characterized the anomalous transmission performance of our metadevice under a $\hat{y}$-polarized normally incident wave. Figures 3.26d and 3.26e show the measured scattered-wave intensities as functions of frequency and deflected angle θ_r/θ_t in transmission and reflection spaces. At the target frequency of 10.6 GHz, the reflected wave is suppressed to a very small portion ($\sim 7\%$, evaluated by integrating power in reflection part). More importantly, almost all transmitted wave is deflected into an anomalous angle with a measured absolute efficiency of $\sim 85\%$ (simulation value: 88%) (Fig. 3.26f), which is consistent with the simulated far-field patterns, as shown Figs. 3.27c,d. Through Figs. 3.26a,e, we can easily observe that out of the designed working frequency, the undesired scattering modes, such as specular reflections and zero-mode transmissions, increase significantly, which limit efficiently the deflecting effect, similar to previously reported metasurfaces [57, 61]. The working bandwidth of our metadevice, defined by the full width at half maximum (FWHM) of efficiency, is about 0.8 GHz (10.2–11 GHz). More importantly, we numerically and experimentally investigate incident angle dependence performance of our metadevice at the desired frequency of 10.6 GHz. Figure 3.28

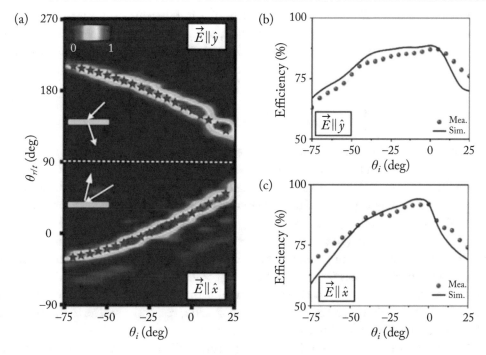

Figure 3.28: Characterizations of our beam deflector under oblique incidence. (a) Measured scattered-field intensities (color map) in reflection (down) and transmission (up) sides vs. incidence angles θ_i and the detecting angles θ_r/θ_t, for our meta-beam-deflector illuminated by $\hat{y}$-polarized (upper panel) and $\hat{x}$-polarized (lower panel) waves. FDT-simulated and measured absolute efficiencies of (b) anomalous reflection and (c) anomalous refraction of our metadevice under oblique illuminations with $\hat{y}$- and $\hat{x}$-polarizations at 10.6 GHz.

shows how the measured normalized intensities of the anomalous reflection/transmission signals vary against incident (θ_i) and detecting angles (θ_r, θ_t), for our beam deflector under illuminations of $\hat{x}$- and $\hat{y}$-polarized microwaves. We can see clearly that the relationships between anomalous reflection/transmission angles and the incident angles well satisfy the generalized Snell's law $\theta_{r/t} = \sin^{-1}(\sin\theta_i + \xi/k_0)$ represented by blue stars. Figures 3.28b and 3.28c further compare the measured and simulated working efficiencies of our device for two functionalities as functions of the incidence angle, from which we find the working efficiencies remain at high values. Excellent agreement between measured and simulated results is noted.

3.4.3   FULL-SPACE FOCUSING LENS

Our findings can stimulate many interesting applications to realize high-performance full-space wavefront manipulation by designing practical phase distributions for reflective and transmissive waves. Here we design a full-space meta-lens, working in reflection and transmission systems for $\hat{x}$- and $\hat{y}$-polarized incident waves, respectively, as shown in Figs. 3.29a,e. The phases ϕ_{xx} and ϕ_{yy} at the position $\vec{r}(x, y)$ of such a metadevice satisfy the following parabolic distributions

$$\begin{cases} \phi_{xx}(x, y) = k_0 \left(\sqrt{F_1^2 + x^2 + y^2} - F_1 \right) \\ \phi_{yy}(x, y) = k_0 \left(\sqrt{F_2^2 + x^2 + y^2} - F_2 \right) \end{cases} \tag{3.10}$$

with F_1 and F_2 being the focal length in reflection and transmission spaces, respectively. Here, we set $F_1 = F_2 = 80$ mm for simplification in the final design, and then ϕ_{xx} and ϕ_{yy} are calculated with the resultant distributions shown in Figs. 3.29c,g which follow well with the parabolic distributions dictated by Eq. (3.10). Then we derive the structural parameters based on the relations of $\phi_{xx} \sim b_1/b_2$ and $\phi_{yy} \sim a$ obtained by FDTD simulations shown in Fig. 3.24. Figures 3.29d and 3.29h depict the corresponding reflection amplitude $|r_{xx}|$ and transmission amplitude $|t_{yy}|$ at each meta-atom of our metasurface. We note that the reflection/transmission amplitudes exhibit high values ($|r_{xx}| > 0.92$ and $|t_{yy}| > 0.85$), implying the high performances of our device. Next, we fabricated an experimental sample according to the design, with its top-view and bottom-view pictures shown in Figs. 3.29b,f, respectively. The sample consists of 14×14 meta-atoms with a total size of 154×154 mm^2.

With the fabricated sample in hand, we experimentally characterize its full-space focusing property. Firstly, we consider the focusing effect operating in reflection space. Shining the sample through a horn antenna with the polarization of $\vec{E} \| \hat{x}$, we measure the electric field distributions at the reflection side by using a monopole antenna (~ 20 mm long). Figure 3.30a displays the measured electric field intensities at both xoz and yoz planes, and the filed is normalized to the maximum value. We can see clearly that the reflected wave has been converged to a focal point at the frequency of $f_0 = 10.6$ GHz. Since the electric field intensity is significantly enhanced at the focal point, we retrieve the measured $|\vec{E}_x|^2$ along z-axis (see Fig. 3.31c) to obtain the maximum value (focal point) at $z = -77$ mm, which agrees well with the theoretical value of $F_1 = 80$ mm. We plot the $Re(\vec{E}_x)$ distributions to straightforwardly observe the wavefront behavior of focusing effect at xoz plane (see Fig. 3.31c). More importantly, we quantitatively evaluate the spot size by the FWHM of the electric-field intensity on the focal plane, with the measured result as 24 mm (see Fig. 3.31d), which demonstrates the desirable focusing effect again. Figure 3.30b shows the measured and FDTD simulated normalized electric field amplitude values at the focal point as a function of frequency. The best performance appears at $f_0 = 10.6$ GHz with the maximum electric field amplitude. The decrement of the electric field is easily understood since the phase distributions ϕ_{xx} do not strictly satisfy the required distributions according to Eq. (3.10). The working bandwidth, defined by the FWHM of $|\vec{E}_x|^2$ at the focal points, is obtained as about 0.8 GHz (~ 10.3–11.1 GHz). We evaluate the working efficiency of our reflective meta-lens in two steps by indirectly measuring the ratio between the power carried by the focal spot and the incident one at the target frequency of 10.6 GHz, with

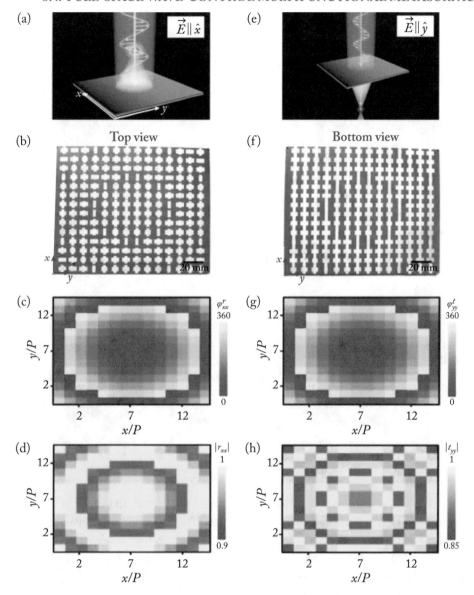

Figure 3.29: Design of a full-space meta-lens. Our metasurface can work as (a) a reflective lens and (b) a transmissive lens when excited by incident waves with polarizations (a) $\vec{E}\|\hat{x}$ and (b) $\vec{E}\|\hat{y}$, respectively. Photographs of our fabricated sample of the meta-lens at (b) top-view and (f) bottom view. (c) and (d) FDTD-simulated ϕ_{xx} and the corresponding $|r_{xx}|$ distributions at each lattice of the designed/fabricated metadevice at the frequency $f_0 = 10.6$ GHz. (g) and (h) FDTD-simulated ϕ_{yy} and the corresponding $|t_{yy}|$ distributions at each lattice of the designed/fabricated metadevice at the frequency of $f_0 = 10.6$ GHz.

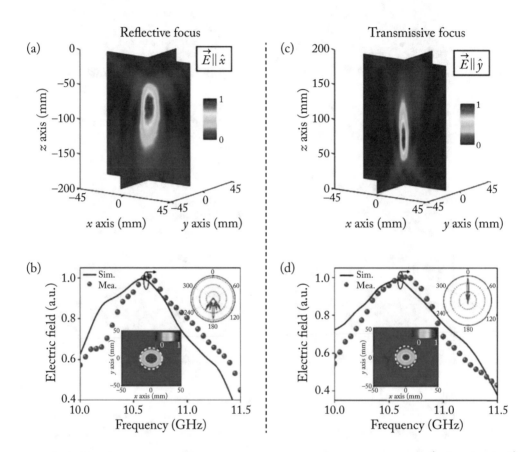

Figure 3.30: Characterizations of our full-space meta-lens. Measured (a) $|\vec{E}_x|^2$ and (c) $|\vec{E}_y|^2$ distributions on both *xoz* and *yoz* planes as the metasurface is illuminated by normally incident (a) $\hat{x}$-polarized and (c) $\hat{y}$-polarized plane waves at $f_0 = 10.6$ GHz. Measured and simulated (b) $\vec{E}_x$-field and (d) $\vec{E}_y$-field amplitude spectra at the focal points under excitations of (b) $\hat{x}$-polarized and (d) $\hat{y}$-polarized plane waves. Inset of (b) depicts the measured $|\vec{E}_x|^2$ distribution at the *xy* plane with $z = -77$ mm and scattered power distribution at $f_0 = 10.6$ GHz. Inset of (d) depicts the measured $|\vec{E}_y|^2$ distribution at the *xy* plane with $z = 78$ mm and scattered power distribution at $f_0 = 10.6$ GHz. All $\vec{E}$-fields are normalized against the maximum value in the spectrum.

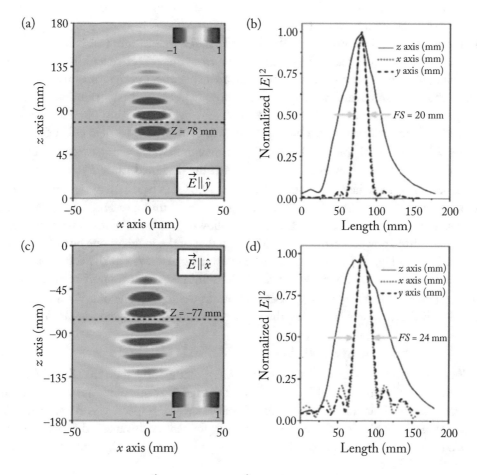

Figure 3.31: Measured (a) $Re(\vec{E}_y)$ and (c) $Re(\vec{E}_x)$ distributions at xoz planes at transmission space and reflection space for our metadevice illuminated by $\hat{y}$- and $\hat{x}$-polarized waves at the working frequency of 10.6 GHz, respectively. Calculated (b) $|\vec{E}_y|^2$ and (d) $|\vec{E}_x|^2$ distributions along three axes from the measured $\vec{E}$-Field distributions for our meta-lens, respectively.

the same approach shown in Section 3.4.2. We experimentally measure the ratio between the reflected power to that of the incident wave for the first step. The measured scattered-wave intensities distributions (see inset to Fig. 3.30b) show that the reflected wave is about 95.8% of the total incident wave. And the missing power partly carried by the dielectric loss ($\sim 1\%$) and partly by the slightly transmitted wave ($\sim 3.2\%$). For the second step, the power taken by the focal point to the reflected power is found to be 94.6%, calculated by integrating the power in the focal point (white dash line) and that in the focal plane of $z = -77$ mm (see inset to Fig. 3.30b). A simple calculation shows that the working efficiency of our reflective lens lies at $\sim 90.6\%$.

Second, we experimentally demonstrate the focusing effect at the transmission space. We can follow the same procedure to measure the electric field intensity distributions at the transmission part on both xoz and yoz planes, respectively, by illuminating our meta-lens with a $\hat{y}$-polarized incident wave through a horn antenna. An obvious electric field enhancement is observed at $z = 78$ mm (focal point), calculated by the maximum $|\vec{E}_x|^2$ along z axis (see Fig. 3.31a). Again, the measured focal length matches well with the designed value $F_2 = 80$ mm. Meanwhile, the best focusing behavior is obtained at $f_0 = 10.6$ GHz based on the spectra of the $|\vec{E}_y|^2$ in Fig. 3.30d, and the measured focal size is evaluated as 20 mm (see Fig. 3.31b). We measure the absolute efficiency of our transmissive lens by following the same method with the reflective lens. The transmitted power is estimated as 89.2% to that of the incident one by the far-field patterns in the inset to Fig. 3.30d, with the missing power carried away by dielectric loss ($\sim 1.2\%$) and reflected wave ($\sim 9.6\%$). Meanwhile, the focal point takes a power portion of 95.3% to the transmitted power, evaluating by $|\vec{E}_y|^2$ distributions at the focal plane of $z = 78$ mm, with the result shown in the inset to Fig. 3.30d. The efficiency is finally calculated as $\sim 85\%$.

3.4.4   FULL-SPACE BIFUNCTIONAL METADEVICE

In previous two parts, we designed two metasurfaces that exhibit the same functionalities for reflected and transmitted waves. Here, we further discuss that our meta-atom can also realize distinct functionalities at transmission and reflection spaces, respectively. As an example, we design a metadevice which combines beam-bending for reflected wave and focusing effect for transmitted wave in a single device. To reach this end, we require ϕ_{xx}^r and ϕ_{yy}^t to satisfy the following distributions

$$\begin{cases} \phi_{xx}^r(x, y) = C_2 + \xi_3 x \\ \phi_{yy}^t(x, y) = k_0 \left(\sqrt{F_3^2 + x^2 + y^2} - F_3 \right) \end{cases} \tag{3.11}$$

with C_2 being a constant, ξ_3 being the phase gradient and F_3 being the focal length. To not lose generality, keeping the center frequency still at $f_0 = 10.6$ GHz, we set $\xi_3 = 0.51k_0$ at $\hat{x}$ polarization, and simultaneously $F_3 = 85$ mm for the transmissive lens at $\hat{y}$ polarization. Aided by the structural map shown in Fig. 3.24, we optimize each meta-atom to satisfy both ϕ_{xx}^r and ϕ_{yy}^t distributions, and then fabricate a microwave sample with the top-view and bottom-view pictures shown in Figs. 3.32b,f. The metadevice contains 15×15 meta-atoms with a total size of 165×165 mm^2. And the corresponding phase (ϕ_{xx}^r and ϕ_{yy}^t) and amplitude ($|r_{xx}|$ and $|t_{yy}|$) distributions are shown in Fig. 3.33. We note that the designed/fabricated phases match well with the theoretical values shown in Eq. (3.11), and reflection/transmission amplitudes keep high values ($|r_{xx}| > 0.92$, $|t_{yy}| > 0.86$), which guarantee the high efficiency of our bifunctional metadevice.

We first perform the experiment to demonstrate the beam bending functionality at the reflection side with an $\hat{x}$-polarized plane wave under normal incidence. The experimental procedures are the same as these of the full-space deflectors in Section 3.4.2. Figure 3.32c depicts the measured scattered angular power distributions in both reflection and transmission spaces. Within a quite broad working frequency interval of 8.4–11.7 GHz, almost all the incident waves are reflected to anomalous angles determined by the generalized Snell's laws (pink symbols in Fig. 3.32c), which can be well reproduced by the FDTD simulations as shown in Fig. 3.34. Obviously, the best performance is found at about 10.6 GHz (as shown in the inset to Fig. 3.32d), demonstrated by the maximum scattered-field power. Moreover, quantitative estimation on the deflection efficiency shows a high absolute efficiency of 88% (simulations: 92%) at this frequency, which is obtained through integration over the angle region of the desirably reflected mode.

We finally examine the focusing functionality at the transmission side under the illumination of a normally incident $\hat{y}$-polarized microwaves. With the same approach shown in Section 3.4.3, we measured the $\vec{E}_y$-field distributions at both xoz and yoz planes. Figure 3.32g clearly shows a good focusing effect at transmission side at 10.6 GHz. Furthermore, we found $|\vec{E}_y|$ at the focal point reaches a maximum value at this frequency, shown in Fig. 3.32h. The focal length is evaluated as 84 mm, in good agreement with the theoretical value of $F = 85$ mm. The FWHM of the focal spot is found as 19 mm, evaluated by the field pattern shown in the inset to Fig. 3.32h. The absolute efficiency of the metalens is experimentally computed in the range of 84.3–85.2%, obtained by the ratio of the beam power carried by the focal area and that of the incident wave.

3.5   SUMMARY

To summarize, this chapter presented the general design strategies and practical realizations of high-performance multifunctional metasurfaces/metadevices working in reflection, transmission and full-space geometries under excitation of linear polarizations. In reflection geometry, two metasurfaces with low polarization cross-talk are designed to realize versatile beam-controls, with the former achieving a pencil beam and four split beams, and the latter achieving an anomalous beam bending and four pencil beams. In transmission geometry, a high-efficiency bifunctional metasurface achieves beam-bending and focusing effects based on the incident polarization. While for the full-space metasurface, we design three metadevices achieving beam bending, focusing, and both functionalities for reflected and transmitted waves.

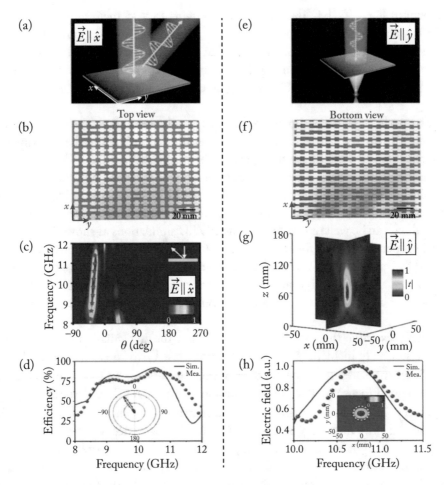

Figure 3.32: Design and characterizations of a full-space bifunctional metadevice. Schematics of (a) the reflective deflector and (e) the transmissive lens of our metadevice under $\hat{x}$ and $\hat{y}$ polarized waves, respectively. (b) Top-view and (f) bottom-view pictures of our fabricated meta-lens. (c) Measured normalized scattered-field intensities (color map) as functions of frequency and the detecting angles θ. (d) FDTD-simulated and measured absolute efficiencies of anomalous reflection of our metadevice shined by an $\hat{x}$-polarized incident wave. Inset shows the measured and FDTD simulated angular distributions of the metasurface, illuminated by an $\hat{x}$-polarized incident wave at $f_0 = 10.6$ GHz. (g) Measured $|\vec{E}_y|^2$ distributions on both xoz and yoz planes in transmission space, as the metasurface is illuminated by a normally incident $\hat{y}$-polarized plane wave at $f_0 = 10.6$ GHz. (h) Measured and simulated $\vec{E}$-field amplitude spectra at the focal points. Insets depict the measured $|\vec{E}_y|^2$ distributions at the xy plane with $z = 84$ mm. All $\vec{E}$-fields are normalized against the maximum value in the spectrum.

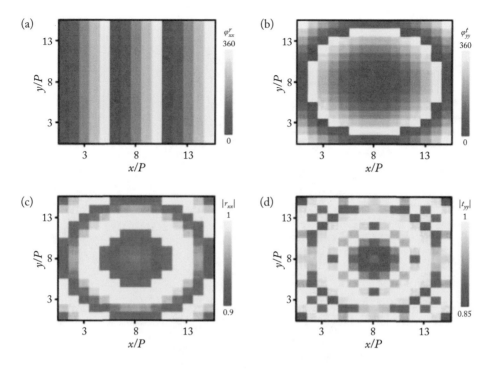

Figure 3.33: FDT-simulated distributions of (a/b) refection/transmission phase and (c/d) amplitude of the designed metadevice at the frequency of 10.6 GHz.

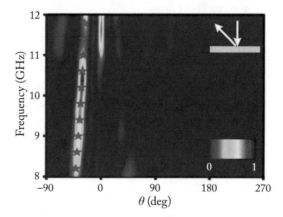

Figure 3.34: FDTD-simulated scattered-field intensities (color map) in full space vs. frequency and the detecting angles, for our bifunctional metasurface under the normally incident illumination with the polarization of $\vec{E}||\hat{x}$.

CHAPTER 4

Multifunctional Metasurfaces/Metadevices Based on Single-Structure Meta-Atoms II: Circular-Polarization Excitations

In the last chapter, we summarized our efforts on achieving multifunctional metadevices under the excitations of EM waves with LPs. In practical applications, however, frequently we need EM devices in responses to EM waves with CPs instead of LPs. Therefore, it is highly desirable to realize metadevices exhibiting distinct wave-control functionalities with respect to CP waves with different handedness. Recently, quite a few multifunctional devices were constructed based on the "merging" approach. However, these devices typically exhibit common shortcomings such as low efficiencies and limited functionalities. In this chapter, we introduce our efforts on realizing high-efficiency multifunctional metadevices working in reflection, transmission and full-space geometries, in responses to CP-wave excitations. To make the story complete and coherent, we start from introducing the general criteria to design 100%-efficiency PB meta-atoms in both reflection and transmission geometries, based on which we next describe our efforts on realizing high-efficiency PB metasurfaces with single functions as benchmark illustrations, and we finally extend the concept to realize PB functional metadevices in controlling both propagating waves and surface waves, in different frequency domains.

4.1 DESIGN PRINCIPLES: PANCHARATNAM–BERRY PHASES

To understand the origin of the Pancharatnam–Berry (PB) phase [154, 155], we need to analyze basic electromagnetic (EM) characteristics of the building block (also called as meta-atoms) of the PB metasurfaces. Consider a homogeneous metasurface consisting of a 2D array of *identi-*

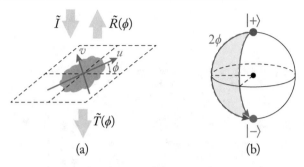

Figure 4.1: (a) Jones matrices of a periodic array of meta-atoms with local axes rotated by an angle φ. (b) The PB phase equals to the area (2φ) surrounded by two curves on the Poincaré sphere with blue and red dashed lines representing the operations of $\hat{\sigma}_-$ and $e^{-i\varphi\hat{\sigma}_3}\hat{\sigma}_-e^{i\varphi\hat{\sigma}_3}$, respectively.

cal subwavelength PB meta-atoms. For simplicity, suppose that such meta-atoms possess two principal axes, we can straightforwardly choose them as the reference coordinates of the system represented by $\{\hat{u}, \hat{v}\}$. Now, the transmission and reflection properties of the PB metasurface can be described by two Jones matrices:

$$T(0) = \begin{pmatrix} t_{uu} & t_{uv} \\ t_{vu} & t_{vv} \end{pmatrix}, \quad R(0) = \begin{pmatrix} r_{uu} & r_{uv} \\ r_{vu} & r_{vv} \end{pmatrix}. \tag{4.1}$$

Let's start by considering the reflection matrix. For the convenience of our discussion, we change linear polarization bases $\{\hat{u}, \hat{v}\}$ to circular polarization ones with the unit vectors $\{\hat{e}_+(0) = (\hat{u} + i\,\hat{v})/\sqrt{2}, \hat{e}_-(0) = (\hat{u} - i\,\hat{v})/\sqrt{2}\}$. The reflection matrix can, therefore, be written as

$$\widetilde{R}(0) = \frac{1}{2}(r_{uu} + r_{vv})\hat{I} + \frac{i}{2}(r_{uv} - r_{vu})\hat{\sigma}_3 + \frac{1}{2}(r_{uu} - r_{vv})\hat{\sigma}_1 + \frac{1}{2}(r_{uv} + r_{vu})\hat{\sigma}_2. \tag{4.2}$$

It is noted that, similar to any other 2×2 matrix, $\widetilde{R}(0)$ has been expanded to the linear combinations of the identity matrix $\hat{I}$ and three Pauli matrices $\{\hat{\sigma}_1, \hat{\sigma}_2, \hat{\sigma}_3\}$.

Next, we consider the PB metasurface with their meta-atoms rotated uniformly by an angle φ with respect to the z-axis (see Fig. 4.1a). With the rotation operator described by $\widetilde{M}(\varphi) = e^{i\varphi\hat{\sigma}_3}$, the reflection matrix of the rotated PB metasurface is written as $\widetilde{R}(\varphi) = \widetilde{M}^\dagger(\varphi)\widetilde{R}(0)\widetilde{M}(\varphi)$ in the CP bases defined in the original (laboratory) frame. According to the commutation relations between Pauli's matrixes, we obtain the following equation:

$$\widetilde{R}(\varphi) = \frac{1}{2}(r_{uu} + r_{vv})\hat{I} + \frac{i}{2}(r_{uv} - r_{vu})\hat{\sigma}_3 + \frac{1}{2}(r_{uu} - r_{vv})$$
$$(e^{-i2\varphi}\hat{\sigma}_+ + e^{i2\varphi}\hat{\sigma}_-) + \frac{i}{2}(r_{uv} + r_{vu})(-e^{-i2\varphi}\hat{\sigma}_+ + e^{i2\varphi}\hat{\sigma}_-). \tag{4.3}$$

Here, $\hat{\sigma}_\pm = (\hat{\sigma}_1 \pm i\hat{\sigma}_2)/2$ represent two spin-flip operators satisfying the following relationship $\hat{\sigma}_\pm|\pm\rangle = 0$ and $\hat{\sigma}_\pm|\pm\rangle = |\pm\rangle$, with $|\pm\rangle$ denoting the spin-up (i.e., $\hat{e}_+$) and spin-down (i.e., $\hat{e}_-$) states, respectively. By replacing $\{r_{uu}, r_{uv}, r_{vu}, r_{vv}\}$ in Eq. (4.3) by $\{t_{uu}, t_{uv}, t_{vu}, t_{vv}\}$, we can obtain the transmission Jones Matrix $\widetilde{T}(\varphi)$ of the rotated PB metasurfaces.

Benefiting from the Jones matrix analysis in Pauli-matrix representation, the underlying physics of Eq. (4.3) becomes extremely clear. It is clear that the first two terms in Eq. (4.3) are the same as those of Eq. (4.2), implying that the reflection properties is the same even after the composite PB meta-atoms are rotated by the angle φ. Meanwhile, the spin state of the corresponding reflection beam is conserved. Since the first two terms in Eq. (4.3) don't provide any phase modulation for impinging wave and thus only contribute to the normal/specular reflection, leading to the degraded performance of all PB devices [156–162]. Conversely, the last two terms will not only flip the spin of the input beam but also provide an additional reflection phase $\pm 2\varphi$, which are the so-called PBs phases with its sign dependent on the spin state of the impinging beam [154, 155]. Such PB phase is totally dispersive, which is completely distinct from the resonance phase adopted in previous metasurfaces composed by different sized meta-atoms [21, 23, 24, 28, 30].

In another line, the nature of the PB phase can be also clarified through a spin-flipping operation on the Poincare sphere, as shown in Fig. 4.1b. For example, the function of operation $\hat{\sigma}_-$ is to flip the spin from $|+\rangle$ to $|-\rangle$, corresponding to an operation of taking the spin state from north to south pole along the red line on the Poincare sphere. However, in the rotating frame, the same operator $\hat{\sigma}_-$ becomes $e^{-i\varphi\hat{\sigma}_3}\hat{\sigma}_- e^{i\varphi\hat{\sigma}_3}$, which give us the same final spin $|-\rangle$ along a different blue loop connecting the two poles on the Poincare's sphere. And the area surrounded by the two different loops (red and blue) on the Poincare sphere corresponds to a phase difference of 2φ (see Fig. 4.1b), according to PB's theory [154, 155].

4.2   PB METASURFACES IN REFLECTION GEOMETRY

Thanks to some unique characteristics, including non-dispersion, easy realization, spin-dependence, PB phases are ideal platforms for manipulating the wavefronts of spin-polarized light [156–162]. Unfortunately, most of them suffer from the problem of low-efficiency that will significantly hamper their practical applications. After obtaining a general Jones matrix described in Eq. (4.3), we can then derive a rigorous design criterion for achieving the 100%-efficiency PB devices [54]. Generally, while the metasurface is normally shined by the linear polarized (LP) wave, the scattered waves generated by different meta-atoms will interfere with each other and form both reflected and transmitted beams traveling along different directions. To achieve a 100%-efficiency PB device, we should prevent the multi-mode generation issue, implying that the meta-atoms should be either totally transparent or totally reflective.

To simplify our design, let's first consider the PB metasurface in reflection mode, in which its building block naturally satisfies $\widetilde{T}(0) \equiv 0$. Therefore, we only need to analyze the theoretical condition constrained on $\widetilde{R}(0)$. Supposing that we form a PB device through combining

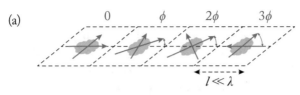

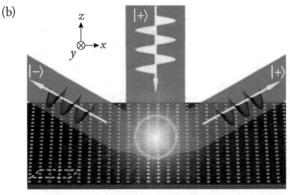

Figure 4.2: (a) A generic PB metasurface formed by meta-atoms with local orientations rotated successively $(0, \varphi, 2\varphi, 3\varphi, \ldots)$. (b) Schematics of 100%-efficiency PSHE realized at our reflective metasurface. A linearly polarized incident beam is split into two spin-polarized reflection beams traveling to two off-normal directions.

the generic PB meta-atoms (of the lattice constant l) with their orientation angles possessing a constant difference of φ, as shown in Fig. 4.2a. According to the previous discussion in Section 4.4.1, the last two terms in Eq. (4.3) will provide a *spin-dependent* linear phase profile for the incident LP beam, generating two opposite off-normal reflection modes with different spins [28, 30]. Since the spin-up and spin-down components of light are spatially separated, such spin-dependent scattering is a kind of photonic spin Hall effect (PSHE) [157–160]. However, the first two terms in Eq. (4.3) provide a *spin-independent* and *constant* reflection phase profile. Therefore, part of the impinging LP beam will be normally reflected to specular direction, which is the key issue degrading the performances of various PB devices [157–160]. As a result, to achieve the 100%-efficiency performance as depicted in Fig. 4.2b, we need to *completely* terminate the normal modes, which means that

$$r_{uu} + r_{vv} = r_{uv} - r_{vu} = 0. \tag{4.4}$$

The derived criterion Eq. (4.4) provides us a clear guideline to design high-efficiency PB meta-atoms in reflection geometry. In addition, the geometric symmetry can further simplify the criterion. For example, supposing that the meta-atom exhibits mirror symmetry, we can obtain $r_{uv} = r_{vu} \equiv 0$, Eq. (4.4) can thus reduce to $r_{uu} = -r_{vv}$. If further neglecting ohmic losses, we have $|r_{uu}| = |r_{vv}| = 1$ in reflection geometry due to energy conservation. Thus, the criterion Eq. (4.4) finally reduces to $\Phi_{vv} - \Phi_{uu} = 180°$, i.e., a constrains on the reflection phase for two cross LP cases. This means that the desired high-efficiency PB meta-atoms should behave effectively as a perfect reflective half wave-plate. Such devices have been realized based on various anisotropic PB meta-atoms in metal-insulator-metal (MIM) configuration working at different frequency domains [12, 13]. Figure 4.3a shows a practical design consisting of a metallic Jerusalem cross structure and a metallic ground plane, separated by a 1.9-mm-thick dielectric film ($\varepsilon = 4.3$) [54]. The bottom metallic mirror not only totally reflect the impinging waves, i.e., ensuring $\widetilde{T}(0) \equiv 0$, but also couple with the top metallic cross structure to create magnetic resonances [163, 164]. The phase of the reflected beam will undergo a full range modulation from 180° to −180° as frequency passes through the magnetic resonance. Via carefully designing the anisotropy of the meta-atoms, the desired phase difference for two polarizations can be achieved. As depicted in Fig. 4.3b, $\Phi_{vv} - \Phi_{uu}$ of this sample can keep at around 180° within a broad frequency band ($\sim$ 10–14 GHz). Based on full-wave optimization, the dispersion cancellation effect between two magnetic resonances with low-quality factors gives rise to this broadband performance. According to Eq. (4.3), we can also evaluate the performance of the PB meta-atoms satisfying mirror symmetry via the formula $|(r_{uu} - r_{vv})/2|^2/R$ (representing the ratio of the anomalous mode relative to the total reflection R), as depicted in Fig. 4.3c.

We can find many different routes to obtain the desired building block satisfying Eq. (4.3). Even for the meta-atom breaking mirror symmetry, the strong off-axis responses can also be utilized to manipulate $\widetilde{R}(0)$. For the lossless case, time-reversal symmetry can also ensure that $r_{uv} = r_{vu}$, implying that the criterion Eq. (4.3) will also reduce to the simple form $r_{uu} + r_{vv} = 0$. Breaking of mirror symmetry makes another extreme solution become possible, i.e., $|r_{uu}| = |r_{vv}| = 0$. It means that, while the system is shined by the EM wave polarized along $\hat{v}$ (or $\hat{u}$), the reflection beam will become the cross-polarized one along $\hat{u}$ (or $\hat{v}$). Numerical optimizations can help us to obtain such asymmetric meta-atom, as depicted in Fig. 4.4a. The measured and simulated reflection spectra of the sample confirm its agreement with Eq. (4.3) within a broad band ($\sim$ 11–14 GHz). We should emphasize that $|r_{uu}| = |r_{vv}| = 0$ is *not* a necessary condition among many different solutions to reach the theoretical criterion.

Let's help the readers to further clarify the working mechanism of such an intriguing asymmetrical meta-atom. We can diagonalize the matrix $\widetilde{R}(0)$ with off-diagonal terms in the original frame $\{\hat{u}, \hat{v}\}$ and then retrieve two *effective* principal axes $\{\hat{u}', \hat{v}'\}$ for our meta-atom. The angle between two coordinate frames shows a dispersive behavior, as depicted in Fig. 4.4c. Quite interestingly, the phase difference between $\Phi_{u'u'}$ and $\Phi_{v'v'}$ defined in the effective frame $\{\hat{u}', \hat{v}'\}$ is also around 180° at the target frequency 12 GHz (see Fig. 4.4d). It means that such asymmetrical meta-atoms share the same working mechanism with the symmetrical one, i.e., behaving effectively as a half wave-plate.

With two different designs of PB meta-atoms, we successfully fabricate the corresponding metasurfaces for achieving high-efficiency PSHE, as illustrated in Figs. 4.5a and 4.5b. Shining the metasurfaces with normally incident LP beams, we adopt, respectively, a left circular polar-

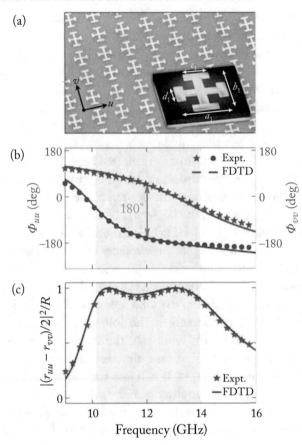

Figure 4.3: Structure and EM properties of the symmetrical meta-atom. (a) Sample picture of an array of symmetrical meta-atoms (sized 7×7 mm^2) with $a_1 = 4$ mm, $b_1 = 5$ mm, $c_1 = 2.2$ mm, $d_1 = 1.3$ mm (see inset). The thickness and width of the metallic wires are 0.05 mm and 0.5 mm, respectively. Spectra of (b) reflection phases (Φ_{uu} and Φ_{vv}) and (c) normalized reflectance $|(r_{uu} - r_{vv})/2|^2/R$ for the sample shown in (a), obtained by experiments (symbols) and FDTD simulations (line). Here, $R = (|r_{uu}|^2 + |r_{vv}|^2 + |r_{uv}|^2 + |r_{vu}|^2)/2$ represents the total reflected energy summing up contributions from all four terms in Eq. (4.2), and the gray region in (b) and (c) indicates the working band of our sample which is determined by the condition of $[|(r_{uu} + r_{vv})/2|^2 + |(r_{uv} - r_{vu})/2|^2]/R < 0.1$.

ization (LCP) antenna and a right circular polarization (RCP) antenna to measure the normalized scattered field angular distributions with $|+\rangle$ (Figs. 4.5c and 4.5g) and $|-\rangle$ (Figs. 4.5b and 4.5f) polarizations. The reference data is obtained by measuring the specular reflection CP signals only with the metasurface replaced by a same-sized metallic mirror. Figure 4.5 demonstrates

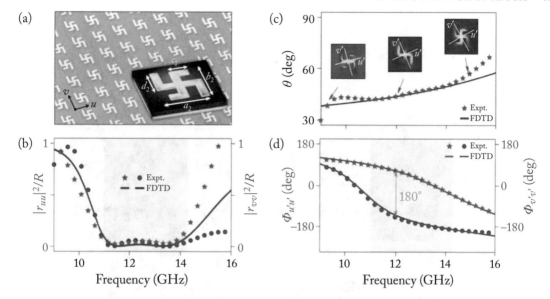

Figure 4.4: Design of the asymmetrical meta-atom. (a) Picture of a sample formed by an array of asymmetrical meta-atoms (sized 7×7 mm^2) with $a_2 = 3.3$ mm, $b_2 = 3.3$ mm, $c_2 = 1.4$ mm, $d_2 = 2.5$ mm (see inset). The thickness and width of the metallic wires are 0.05 mm and 0.5 mm, respectively. (b) Measured and simulated spectra of $|r_{uu}|^2/R$ and $|r_{vv}|^2/R$ for the sample shown in (a). (c) Orientation angle θ of the effective principal axes $\{\hat{u}', \hat{v}'\}$ for our asymmetrical meta-atom with respect to its original coordinate $\{\hat{u}, \hat{v}\}$. Insets show the current distributions on the metallic wires in our meta-atom, excited by normally incident electromagnetic waves polarized $\vec{E}\|\hat{v}$ at frequencies 9 GHz, 12 GHz, and 15 GHz, respectively. (d) Measured and FDTD simulated spectra of reflection phases ($\Phi_{u'u'}$, $\Phi_{v'v'}$) for the sample shown in (a).

that both metasurfaces can deflect the spin-up and spin-down components of the impinging LP beam to two distinct directions within a broad frequency band, verifying the giant PSHE. More importantly, the specular (or normal) refection modes almost disappear within the working frequency band, implying the high efficiency of our devices. Via integrating the scattered field power carried by the anomalous modes, we can retrieve quantitatively the PSHE efficiencies for two metasurfaces, as depicted in Figs. 4.5d and 4.5h. The maximum value can go beyond 90% around the central working frequency. Outside the working frequency band, specular reflection modes become strong, degrading the efficiency of PSHE.

The PSHE achieved by proposed PB metasurfaces is governed by the generalized Snell's law [21, 23, 24, 28, 30]

$$\theta_r^{\pm} = \sin^{-1}\left(\sin\theta_i + \xi^{\pm}/k_0\right), \tag{4.5}$$

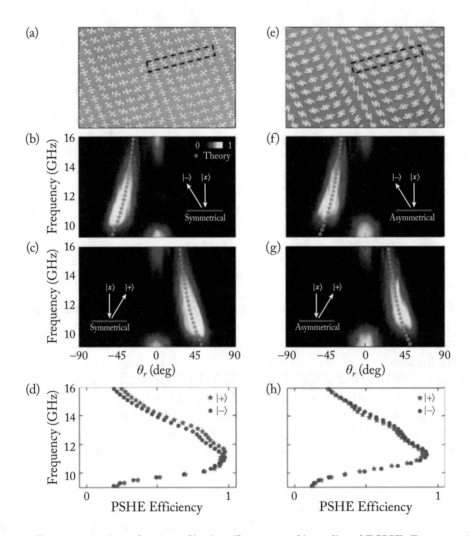

Figure 4.5: Experimental verification of high-efficiency and broadband PSHE. Pictures of fabricated metasurfaces (both sized 504×504 mm^2) formed by (a) symmetrical and (e) asymmetrical meta-atoms. Measured normalized scattered-field intensities (color map) vs. frequency and detecting angle for two metasurfaces illuminated by normally incident linearly polarized beams, with receivers chosen as a circularly polarized antenna with polarization (b, f) $|+\rangle$ and (c, g) $|-\rangle$, respectively. (d, h) PSHE efficiencies vs. frequency for two metasurfaces, obtained by analyzing the experimental data in (b, c, f, g). Here dotted lines in (b, c, f, g) are calculated by Eq. (4.4) under normal-incidence condition. Regions surrounded by black dashed lines in (a, e) represent the super-cells of two samples.

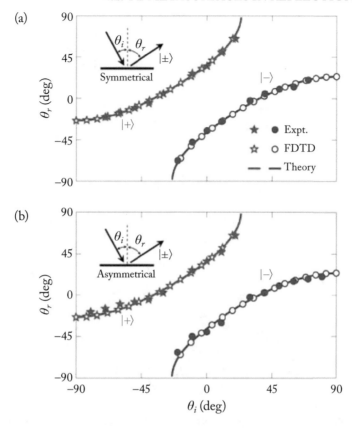

Figure 4.6: Angle of the anomalously reflected beam θ_r vs. the incident angle θ_i for metasurfaces with (a) symmetrical and (b) asymmetrical meta-atoms under illuminations of spin-polarized waves, obtained from experiments (solid symbols), FDTD simulations (open symbols), and theory (Eq. (4.4), lines) at frequency 12 GHz.

where $\theta_r^{\pm}$ is the anomalous reflection angle for different spins, θ_i is the incident angle, and $k_0 = \omega/c$ is the wave-vector in the vacuum. It is clear that the significance of PSHE, i.e., the spin-dependent deflection angle $\theta_r^{\pm}$, is well controlled by the phase gradient $\xi^{\pm} = \pm 2\varphi/l$. Equation (4.5) can predict the deflection angle of the anomalous mode generated by two fabricated samples (dotted lines in Figs. 4.5b,c,f,g), showing good agreement with the measurement results (colormaps). Meanwhile, we also measured the relations of $\theta_r^{\pm} \sim \theta_i$ for two spin-polarized reflection modes at a fixed frequency of 12 GHz, as depicted in Fig. 4.6. The measurements again match well with the simulations and theoretical predictions.

Spin-dependent light manipulations based on PB metasurface can stimulate many interesting applications. For example, Fig. 4.7a depicts the schematic of a new kind of efficient and

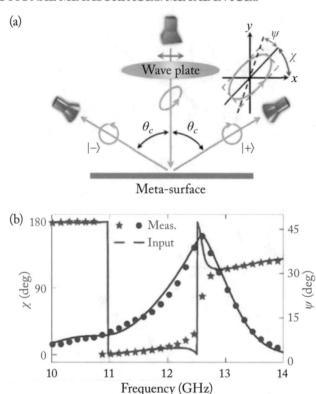

Figure 4.7: PB metasurface based polarization detector. (a) Schematics of our polarization-detecting experiments. Inset depicts a general elliptic polarization state defined by the orientation angle χ and the degree of ellipticity ψ. (b) Polarization states of the input beam measured by our experiments (symbols), compared with the input values (lines).

broadband polarization detector based on the proposed metasurfaces. Here, the proposed PB metasurface can efficiently deflect two spin components of an impinging wave with unknown polarization to two off-normal directions, which are measured simultaneously by two detectors (including the amplitudes and phases) for retrieving its original polarization state. Figure 4.7b shows that the measured polarization states of a well-known input beam, characterized by two parameters (the orientation angle and the ellipticity angle ψ, see inset to Fig. 4.7a), showing excellent agreement with its input polarization. Compared to conventional polarization detectors, our scheme is more efficient and robust.

The criterion Eq. (4.3) can guide us to achieve various kinds of PB devices with high efficiency. For instance, while PB metasurfaces were also proposed for spin-controlled SPP excitations [158, 159], their working efficiencies are extremely low ($< 10\%$). We theoretically analyzed

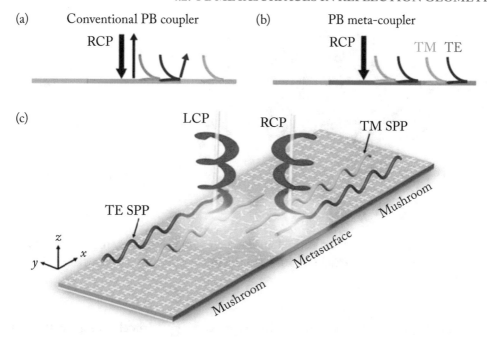

(a) Conventional PB coupler

RCP

(b) PB meta-coupler

RCP   TM TE

(c)

LCP   RCP   TM SPP

TE SPP

Mushroom

Metasurface

Mushroom

z

y x

Figure 4.8: New concept for realizing the high-efficiency and direction-controllable SPP meta-coupler. (a) Schematic of conventional low-efficiency PB meta-coupler. (b) Proposed PB meta-coupler consisting of a high-efficiency PB metasurface and artificial plasmonic metals with previous two issues well solved. Here, the new PB meta-coupler can convert the input CP wave into driven SWs with nearly 100% efficiency, and the artificial metals can guide out both TE and TM polarized SWs to corresponding SPPs with their wavevectors perfectly matching. (c) Schematics of the realistically designed PB meta-coupler illuminated by the LCP or RCP waves, producing high-efficiency chirality-controlled SPPs.

the inherent issues limiting the performance of these PB systems: (i) their building blocks do not satisfy the 100%-efficiency criterion, i.e., Eq. (4.4), so that normal modes inevitably exist and thus degrade their performances; and (ii) while the incident CP light contains both transverse-electric (TE) and transverse-magnetic (TM) polarized components, the SPP modes excited on the connecting metals are TM polarization. Such a polarization mismatching issue further decreases the surface plasmon conversion efficiency (see Fig. 4.8a). Recently, we propose a new scheme to design high-efficiency and chirality-modulated surface plasmon meta-coupler, with these two issues well resolved [165]. First, the adopted meta-atoms for constructing PB metasurfaces should satisfy the 100%-efficiency criterion Eq. (4.3), in order to terminate all of the normal reflection modes. Second, an artificial plasmonic metal is also carefully designed to support both TE and TM polarized surface plasmons, in order to guide out nearly all the energies

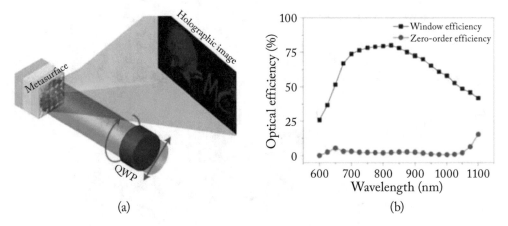

(a)   (b)

Figure 4.9: (a) Illustration of the reflective nanorod based high-efficiency metahologram. (b) Experimentally obtained optical efficiency for both the image and the zeroth-order (normal mode) beam [44].

carried by the input CP light (see Fig. 4.8b). With such appropriately designed PB metasurface and plasmonic metal, a high-efficiency and direction-controllable SPP excitation should in principle be achieved via integrating them together, as shown in Fig. 4.8c. Based on this scheme, we successfully constructed a high-efficiency PB meta-coupler for microwave frequencies and experimentally demonstrated that both TE and TM polarized SWs are efficiently guided out as the eigen SPPs. The adopted PB meta-atom (see Fig. 4.8c), composed by a metallic H-structure and a flat metal mirror separated by a 2-mm-thick dielectric spacer, also possesses the carefully designed structural anisotropy and thus behave like a half wave-plate. Far-field measurements finally demonstrated that the realized coupler exhibits a very high conversion efficiency (78%), which can be further enhanced based on careful optimizations.

The theoretical criterion is so general that can be applied in various frequency domains. For example, Zhang's group independently discover the design strategy and experimentally demonstrate a high-efficiency meta-hologram in optical frequency [44]. The adopted meta-atom is also a MIM structure with the top resonator being a simple nano-rod, which can function as a half wave-plate from 550–1000 nm wavelength (see Fig. 4.9b). The authors construct a metahologram encoding the image of Einstein's portrait (see Fig. 4.9a), with the orientation angles of composite PB meta-atoms determined by CGH-calculated phase distribution. Experiments verified that the maximum efficiency can reach 80% at 825 nm wavelength.

Due to the limited space of this book, we cannot provide too many discussions on the PB metasurfaces in reflection geometry, which is indeed a hot topic in the community [36, 56, 166–170].

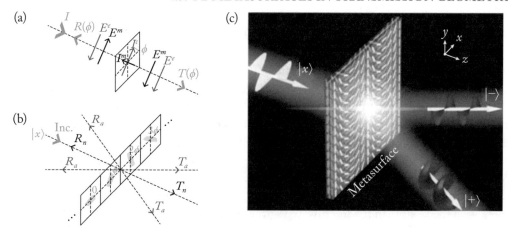

Figure 4.10: (a) Schematics of a φ-orientated meta-atom with transmission/reflection characteristics described by two Jones matrices $\boldsymbol{T}$ and $\boldsymbol{R}$, and the E fields radiated from the electric and magnetic currents ($\vec{J}^e$ and $\vec{J}^m$) generated on the meta-atom. (b) Anomalous/normal transmissions and reflections (with efficiencies of T_a, R_a, T_n, R_n) generated by a generic PB metasurface consisting of meta-atoms with spatially varying orientation angles $(0, \varphi, 2\varphi, 3\varphi, \ldots)$, under the illumination of a linearly polarized incident beam. (c) Schematics of the 100% efficiency PSHE achieved by a PB metasurface satisfying Eq. (4.7), with $|x\rangle, |+\rangle, |-\rangle$ representing beams with linear polarization, left circular polarization, and right circular polarization, respectively.

4.3   PB METASURFACES IN TRANSMISSION GEOMETRY

High-efficiency PB metasurfaces in transmission geometry are more desirable for real applications, but are more difficult to realize experimentally, especially in ultra-thin metasurfaces [171, 172]. To solve such a problem, we symmetrically analyze the PSHE efficiency of a generic PB metasurface as shown in Fig. 4.10, based on the Jones Matrix method [54]. As the PB metasurface is illuminated with a spin-polarized light, four light beams will be generated as anomalous or normal transmission or reflection modes, respectively (Fig. 4.10b). Based on recently established theory in [173], the efficiencies of such four beam are

$$T_n \equiv |(t_{uu} + t_{vv})|^2/4, \quad R_n \equiv |(r_{uu} + r_{vv})|^2/4,$$
$$T_a \equiv |(t_{uu} - t_{vv})|^2/4, \quad R_a \equiv |(r_{uu} - r_{vv})|^2/4, \tag{4.6}$$

where the subscripts "a" and "n" stand for anomalous and normal modes, respectively. Therefore, a 100%-efficiency transmissive PB metasurface can be realized, as long as the constitutional meta-atom exhibit Jones matrix element that satisfies the following condition:

$$r_{uu} = r_{vv} = 0, \quad |t_{uu}| = |t_{vv}| = 1, \quad \arg(t_{uu}) = \arg(t_{vv}) \pm \pi. \tag{4.7}$$

Such a condition means the meta-atom should function as a half-wave plate (HWP) with 100% transmission. However, it can never be fulfilled by a single-layer PB meta-atom with only electrical response, since an electric dipole radiates symmetrically to both sides, and we thus cannot independently control the transmitted and reflected beams. We argue that the desired meta-atom having both electrical and magnetic responses, as shown in Fig. 4.10a, can provide more degree of freedom to alter the transmission and reflection properties, thanks to obtained constructive interference at the transmission side and the destructive interference in reflection side. Such arguments are in the same spirit as previous high-transmission arguments in different scenarios, such as Huygens metasurfaces [24, 174–176] and dielectric metasurfaces [177, 178]. Further adding appropriate anisotropy into the meta-atom design, a transmissive HWP can be achieved. Inspired by [10, 31, 179], we adopted the ABA trilayer system to construct a 100% efficiency PB meta-atom in the microwave regime as shown in Fig. 4.11a, in which the magnetic response is induced by the coupling between two metallic layers. The meta-layer A is consisting of an array of metallic bars, while the meta-layer B is composed of a holey metallic film coupled with a metallic bar, and two dielectric spacers are used to separate them. The thickness of the ABA meta-atom is about 4 mm ($\sim \lambda/8$). The measured and simulated results shown in Figs. 4.11c and 4.11e illustrate well the ABA meta-atom is optically transparent for both polarizations with the transmission phase different $\sim 180°$ in the working frequency window covering 10.1–10.9 GHz. By putting the Jones matrix parameters (Figs. 4.12c and 4.12e) into Eq. (4.6), we can obtain the predicted efficiencies of four beams (T_a, T_n, R_a, R_n), for a PB meta-surface made by such a meta-atom. Figures 4.12d and 4.12f illustrate that nearly 100% working efficiency of the anomalous transmission T_a can be achieved within 10.1–10.9 GHz, with all other energy channels suppressed almost completely ($T_n \approx R_a \approx R_n \approx 0$).

With such a building block with ideal functionality of HWP in transmission geometry in hand, we successfully designed and fabricated three PB metasurfaces and experimentally demonstrated that they can realize PSHE, vortex beam generation and Bessel beam generation with very high efficiencies in the microwave regime.

We first fabricated a PB metasurface with inter-particle rotation angle $\varphi = 30°$ (see Fig. 4.12a) to realize the PSHE effect and verify our theoretical prediction presented in Fig. 4.12. To experimentally characterize its PSHE performance [173], we shine an LP microwave normally onto the PB metasurface and measured the scattered EM wave with LCP and RCP utilizing corresponding horn antennas (see insets to Fig. 4.12). Figures 4.12b–f, depicting the measured spin-dependent scattering pattern of designed metasurfaces at both transmission and reflection side, illustrate well that, within the working frequency range, only the anomalous transmission mode survive with all other three modes largely suppressed. Outside the working frequency band, the anomalous transmission efficiency gradually decreases. The efficiency of all four modes can be evaluated through the integration over the angle regions of the different modes. Consistent with the Jones matrix analysis, the measured PSHE efficiency (says, anomalous transmission mode) of such PB metasurface reach as high as 91% at 10.5 GHz (see

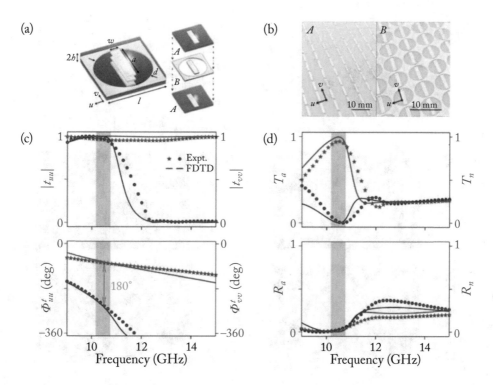

Figure 4.11: (a) Schematics of the ABA meta-atom and individual A-structure and B-structure. Dielectric spacers with thickness $h = 2$ mm and $\varepsilon = 4.6$ are adopted to separate three metallic structures to form the final ABA meta-atom. (b) Pictures of fabricated layer A (left) and layer B (right). Measured and FDTD simulated spectra of transmission (c) amplitude and (e) phase for a metasurface consisting of a periodic array of ABA meta-atoms (with periodicity 7 mm × 7 mm). Efficiencies of (d) anomalous/normal transmissions and (f) anomalous/normal reflections for a PB metasurface constructed with the ABA meta-atoms studied in (c, e), calculated with Eq. (4.6) based on measured/simulated transmission/reflection characteristics of the meta-atom. Here, $a = 5.8$ mm, $w = 1.1$ mm, and $d = 6.5$ mm.

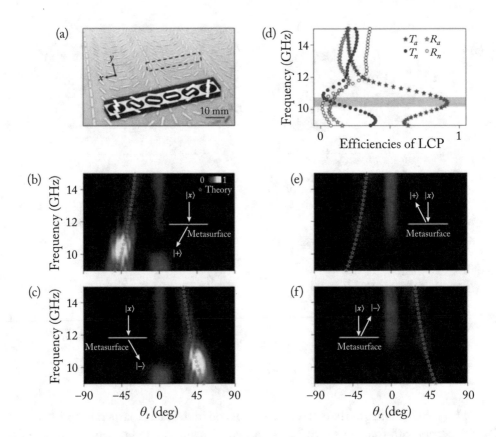

Figure 4.12: (a) Picture of fabricated PB metasurface based on the *ABA* meta-atoms studied in Fig. 4.11. Measured scattered-field intensities (color map) in (b, c) transmission and (e, f) reflection sides as a function of detecting angle and frequency, shined by a normally incident LP beams. The receivers are circularly polarized antennas with polarization (b, f) LCP and (c, e) RCP, respectively. (d) Absolute efficiencies of four modes as a function of frequency for the PB metasurface, obtained by integrations over appropriate angle regions of different modes based on the experimental data. Blue stars in (b, c, e, f) represent the analytical results based on the generalized Snell's law.

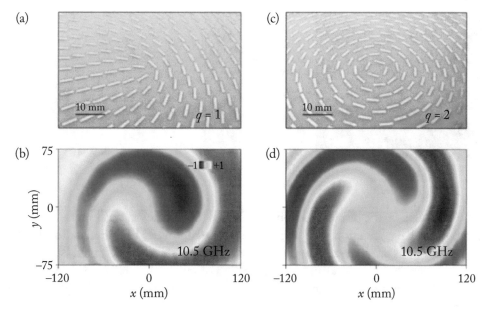

Figure 4.13: (a), (c) Pictures of vortex-beam generators with topological charge $q = 1$ and $q = 2$, realized with the *ABA* meta-atom, respectively. (b), (d) Measured $Re(E_x)$ distributions on an x-y plane 50 cm below such two metasurfaces with illumination of a normally incident RCP waves at 10.5 GHz.

Fig. 4.12d). In addition, the spin-dependent deflection angles of the scattered wave of the PB metasurfaces at different frequencies, governed by the generalized Snell law [21, 23, 24, 28, 30], are in good agreement with experimental results.

Our PB meta-atom can be used to realize other high-performance functional metadevices in transmission geometry [173]. As an example, we designed and fabricated vortex-beam generators with a collection of meta-atoms with at that at the point (x, y) exhibiting an orientation angle φ satisfying $\varphi = (q/2)\tan(y/x)$, as shown in Figs. 4.13a and 4.13c. Here, q is the topological charge. To characterize the properties of generated vortex beam, we shine these devices with normally incident RCP beams and measure the $Re(E_x)$ distributions of the transmitted EM wave. Figures 4.13b and 4.13d show that *pure* vortex beams with $q = 1, 2$ at 10.5 GHz have been generated, thanks to interferences among anomalous-mode waves passing through meta-atoms at different positions.

We further designed and fabricated a meta-axicon with such a high-efficiency PB meta-atom (see Fig. 4.14a), which can generate CP Bessel Beams (BB) in the microwave regime [180]. Since such PB metasurfaces should exhibit the transmission phase distribution $\varphi(x, y) = k_{//}\sqrt{x^2 + y^2}$, we arranged the PB meta-atoms in a bi-dimensional plane, with the meta-atom placed at point (x, y) having an orientation angle $\alpha = \varphi(x, y)/2$. Therefore, when shining such

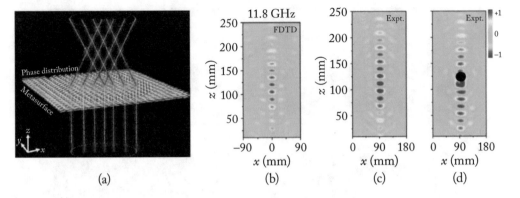

Figure 4.14: (a) Schematic of meta-axicon based on a 100%-efficiency PB meta-atom in ABA configuration for BB generation with high efficiencies. (b), (c) FDTD simulated and measured Ey field distributions on the x-z-plane at the transmission side of the designed BB generator at 11.8 GHz, respectively. (d) Experimental verification of self-healing effect of the generated BB based on Ey distribution of the meta-axicon with the presence of a metallic sphere (diameter $D = 15$ mm).

PB metasurface with LCP light, the transmitted beam will be RCP beam with an additional phase factor $e^{i2\alpha}$. Figures 4.14b and 4.14c depict the FDTD-simulated and measured Ey field distribution on the x-z plane (for $y = 0$) at 11.8 GHz, which well illustrates the non-diffraction properties of the generated BBs. The self-healing property of such BBs is also experimentally demonstrated by placing a metallic sphere on its propagation trajectory. Comparing Figs. 4.14c and 4.14d, while the BB undergoes some scattering due to the presence of the metallic sphere, some waves can pass through the scatter and reconstruct the desired BB field patter in the far-field. In addition, the self-healing effect is insensitive to the shape and position of the scatter. The measured working efficiency of such CP BBs generator reaches 91% at 12 GHz, well consistent with FDTD simulations ($\sim$ 92%).

Our design principle of such PB metasurfaces can be easily extended to high-frequency regime, which provides us a promising platform to efficiently control the CP light in different frequency ranges. Here we present one typical example of CP BB generation in THz regime based on our transmissive PB metasurfaces design [181], since the efficient manipulation of CP THz wave is highly desired due to both curiosities in fundamental science and immense techni-cal demands in applications (i.e., biological sensing, THz telecommunication). According to the design strategy, we first designed a PB meta-atom functioning as a THz HWP in transmission geometry based on a freestanding anisotropic *ABA* configuration. As shown in Fig. 4.15a, layer *A* in our meta-atom is a U-shaped metallic resonator, layer *B* is a metallic plate with holes cou-pled with the same U-shaped resonator, and two polyimide spacers are employed to separate the three metallic layers. The interlayer coupling introduced effective magnetic currents makes the

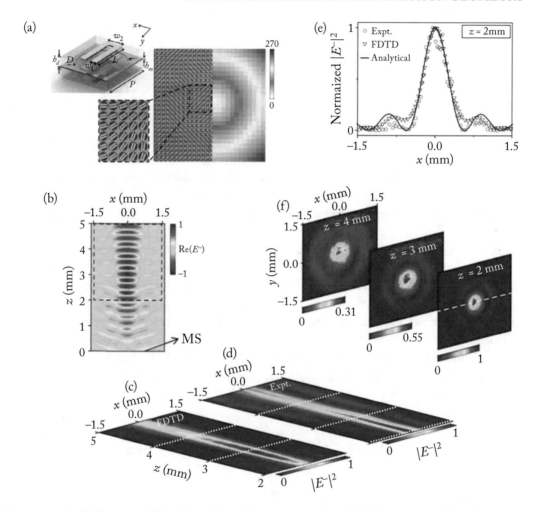

Figure 4.15: Background-free Bessel beam generation for CP THz waves. (a) Schematic of meta-atom design and picture of part of a fabricated CP Bessel beam generator which is a PB metasurface (left panel) designed based on a particular transmission-phase distribution (right panel). (b) FDTD simulated $Re(E^-)$ distribution on the x-z-plane with $y = 0$ mm for our metasurface (placed at $z = 0$ mm) shined by a normally incident x-polarized THz beam at 0.66 THz. (c) FDTD simulated and (d) z-scan measured $|E^-|^2$ distributions inside the area surrounded by black dashed lines in (b), under exactly the same conditions as in (b). (f) Measured $|E^-|^2$ distributions on x-y planes with $z = 2, 3,$ and 4 mm, respectively. (e) Normalized $|E^-|^2 \sim x$ distributions along the line with $z = 2$ mm and $y = 0$ mm, obtained by experiment (red circles), FDTD simulations (blue triangles) and theoretical formula for zero-order BB (solid line).

meta-atom exhibit high transmission property, meanwhile, the U-shape provides more free-dom to generate lateral anisotropy for EM responses of different polarizations. After experi-mentally verified the performance of such meta-atom as transmissive HWP, we then employ such meta-atom as the building block to construct a PB metasurface that can efficiently gen-erate high-performance CP THz BBs without normal-mode background interference. Similar to the idea shown in [180], such ultra-thin PB metadevice with appropriate phase distribution (Fig. 4.15a), can function as an axicon to bend an incident CP wave to the desired angle to the transmissive side, thus generating the desired CP BB. Figure 4.15b depicts the FDTD simu-lated $Re(E^-)$ (i.e., RCP field component) distribution in x-y-plane on the transmission side of our PB device, with illumination of an x-polarized normally incident beam at 0.66 THz, which clearly illustrates the non-diffracting property of generated RCP BB. We experimen-tally characterized the performance of such CP BB generator with a THz digital holographic imaging system [182, 183], and demonstrated the non-diffracting features of the generated BB with the FDTD simulated and measured intensity profile for the transmitted RCP beam in x-z-plane at 0.66 THz (see Figs. 4.15c and 4.15d), respectively.) Figure 4.15f depicts the measured intensity distribution of the RCP component in different x-y planes at the different longitudinal position, showing the rotationally invariant symmetries with strengths which decay quickly from the center. We also compare the intensity distribution along x axis, obtained by the FDTD simulations, experiments, theoretical analysis for the designed zero-order BBs. All the measured/simulated results illustrate that the generated BB is not affected by the interference from the normal transmitted mode background, thanks to the relatively high efficiency of the constitutive PB meta-atom.

4.4   PB METASURFACES IN FULL-SPACE GEOMETRY

In this section, we explore to design full-space bifunctional metasurfaces under circular-polarization excitations within subwavelength thickness. Different functionalities are realized at both sides of metasurfaces relying on incident chirality [140]. For demonstration, we designed two metadevices with the first working as a bifunctional metalens (converging the transmissive wave for LCP incidence and diverging the reflective wave for RCP incidence), and the second behaving as a CP beam splitter. Both metasurfaces exhibit very high efficiencies better than 88%, indicating many interesting applications to achieve full-space bifunctional metasurfaces at microwave frequency.

4.4.1   CONCEPT AND META-ATOM DESIGN
We start by analyzing the working principle of full-space metasurfaces under circular-polarization excitations. As discussed in the last two sections [54, 173], the EM characteristics

of the meta-atoms can be described by two Jones matrices

$$R_{lin} = \begin{pmatrix} r_{xx} & r_{xy} \\ r_{yx} & r_{yy} \end{pmatrix} \quad \text{and} \quad T_{lin} = \begin{pmatrix} t_{xx} & t_{xy} \\ t_{yx} & t_{yy} \end{pmatrix},$$

with $r_{xx}, r_{xy}, r_{yx},$ and $r_{yy}/t_{xx}, t_{xy}, t_{yx},$ and t_{yy} being the reflection/transmission coefficients for $\hat{x}$- and $\hat{y}$-polarized waves. While under the circular base, the transmission and reflection matrices can be computed as

$$T_{cp} = \Lambda^{-1} T \Lambda = \begin{pmatrix} t_{++} & t_{+-} \\ t_{-+} & t_{--} \end{pmatrix} \quad \text{and} \quad R_{cp} = \Lambda^{-1} R \Lambda = \begin{pmatrix} r_{++} & r_{+-} \\ r_{-+} & r_{--} \end{pmatrix},$$

respectively, with

$$\Lambda = \frac{1}{\sqrt{2}} \begin{pmatrix} 1 & 1 \\ j & -j \end{pmatrix},$$

the subscripts $+$ and $-$ denoting the circularly polarized waves with clockwise and counterclockwise. Thus, r_{-+} and r_{+-} are co-polarizations for the reflected waves and t_{++} and t_{--} for the transmitted waves since their wavevectors are in opposite directions. Ignoring the material loss, we can obtain $|r_{++}|^2 + |r_{-+}|^2 + |t_{++}|^2 + |t_{-+}|^2 = 1$ and $|r_{--}|^2 + |r_{+-}|^2 + |t_{--}|^2 + |t_{+-}|^2 = 1$ due to the energy conservation. We have discussed the totally reflective system (satisfying $|r_{++}|^2 = 1$ (or $|r_{-+}|^2 = 1$) and $|r_{--}|^2 = 1$ (or $|r_{+-}|^2 = 1$) [54], as shown in Fig. 4.16a) and totally transmissive case ($|t_{++}|^2 = 1$ (or $|t_{-+}|^2 = 1$) and $|t_{--}|^2 = 1$ (or $|t_{+-}|^2 = 1$), as depicted in Fig. 4.16b) in last two sections [173]. In this section, we discussed how to control both the transmitted wave and reflected wave, requiring our meta-atom to be perfectly reflective for the "$+$" polarized incidence, and perfectly transparent for the "$-$" polarized incidence. Here, two conditions satisfy this criterion. First, when the Jones matrices are

$$R_{lin} = \frac{1}{2} \begin{pmatrix} 1 & -j \\ -j & -1 \end{pmatrix} \quad \text{and} \quad T_{lin} = \frac{1}{2} \begin{pmatrix} 1 & j \\ j & -1 \end{pmatrix}$$

in a linear system, with corresponding

$$R_{cp} = \begin{pmatrix} 0 & 0 \\ 1 & 0 \end{pmatrix} \quad \text{and} \quad T_{cp} = \begin{pmatrix} 0 & 1 \\ 0 & 0 \end{pmatrix}$$

under a circular base. We note that both transmitted and reflected waves carry PB operations under this condition. And the reflected wave has the same helicity with the incidence, while the transmissive wave has an inverse polarization, as shown in Fig. 4.16c. Second, when the Jones matrices are

$$R_{lin} = \frac{1}{2} \begin{pmatrix} 1 & -j \\ -j & -1 \end{pmatrix} \quad \text{and} \quad T_{lin} = \frac{1}{2} \begin{pmatrix} 1 & j \\ -j & 1 \end{pmatrix}$$

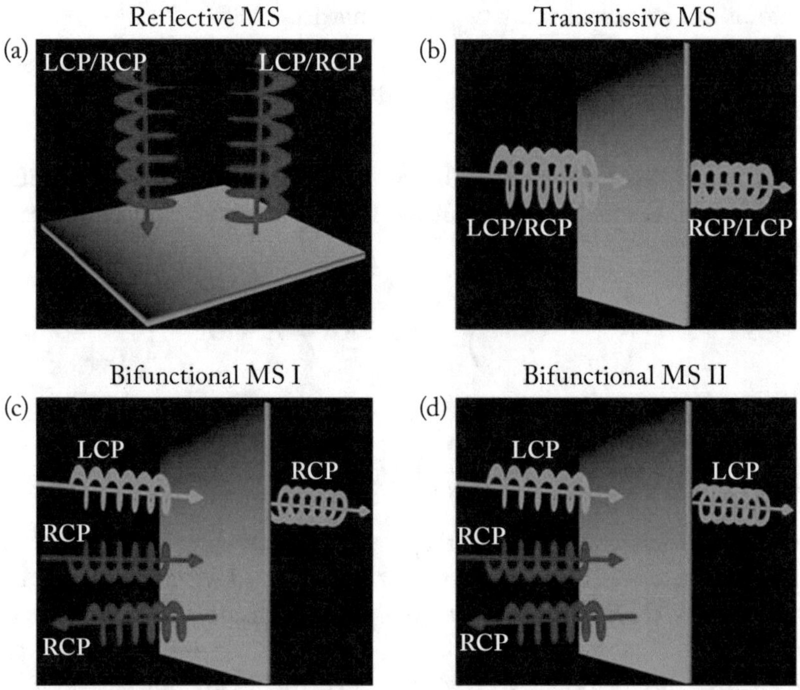

Reflective MS   Transmissive MS

Figure 4.16: Working mechanism of full-space PB metasurfaces. Schematics of (a) reflective PB metasurface and (b) transmissive metasurface. (c) Full-space metasurface I can reflect the RCP wave and preserve its handedness, while transmit the LCP incidence and reverse its handedness. (d) Full-space metasurface II can reflect the RCP wave and transmit the LCP incidence while keeping both the handednesses of the incident waves.

under linear base, corresponding to

$$R_{cp} = \begin{pmatrix} 0 & 0 \\ 1 & 0 \end{pmatrix} \quad \text{and} \quad T_{cp} = \begin{pmatrix} 0 & 0 \\ 0 & 1 \end{pmatrix}$$

in a circular system. In this case, as shown in Fig. 4.16d, the PB operation exists only in reflected wave, and the polarizations for both reflected and transmitted waves remain unchanged. Therefore, we can realize full-space wavefront control of CP waves under these two conditions.

Then, we consider how to realize those Jones matrices in realistic structures. The first meta-atom is composed of three layers of composite resonators separated by two F4B substrates with $h = 2$ mm, with its topology shown in Fig. 4.17a. The working frequency is chosen as $f_0 = 11$ GHz. We can obtain the design principle easily based on the required Jones matrices R_{lin} and T_{lin}, which can be built by the matrices of quarter-wave plates at first and third terms and a matrix of linear polarizer at the second term. In our designed meta-atom, the anisotropically

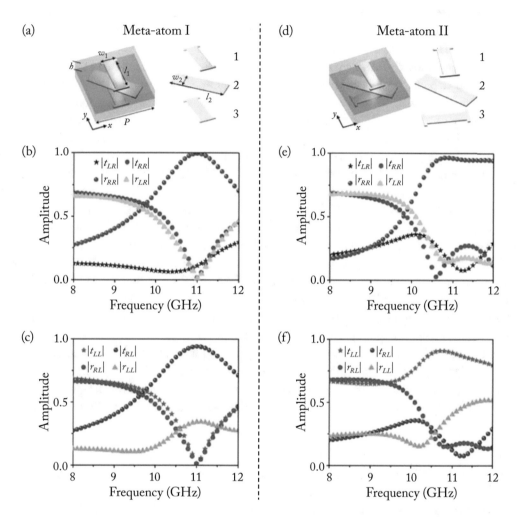

Figure 4.17: EM characterizations of the proposed full-space meta-atoms. Topology of the (a) meta-atom I and (d) meta-atom II, which are composed of three metallic layers separated by two F4B spacers ($\varepsilon_r = 2.65 + 0.01i$, $h = 2$ mm) with parameters fixed as $p = 11$ mm, $w_1 = w_2 = 3$ mm, $l_1 = 7$ mm, and $l_2 = 9.6$ mm. The FDTD simulated transmission/reflection spectra for the metasurfaces constructed by periodic arrays of meta-atoms (b, c) I and (e, f) II, under excitations of (b, e) RCP and (c, f) LCP waves.

metallic "I" resonator can be regarded as a quarter-wave plate, while the middle oblique bar plays as a role of a linear polarizer. In meta-atom II, the Jones matrices can be calculated by replacing one of the quarter-wave plates with its conjugated part, which is realized by rotating the lower "I" structure as 90° and keeping the other resonators unchanged, with the schematic of meta-atom II shown in Fig. 4.17d.

Next, we study the EM characterizations of both meta-atoms via FDTD simulations. Figure 4.17b depicts the reflection and transmission amplitude spectra under excitation of a normally incident RCP wave. We can see that the incident RCP wave is totally reflected at the frequency of $f_0 = 11$ GHz, while a completely transparent window ($|t_{RL}| \approx 1$) appears for an LCP incident wave, as shown in Fig. 4.17c. By using the same approach, we can obtain the amplitude spectra of the meta-atom II, as shown in Figs. 4.17e and 4.17f. This meta-atom can simultaneously achieve high reflection for RCP incidence and high transmission for LCP incidence. More importantly, the helicity keeps unchanged for incident waves with different polarizations in this case, which is of importance for designing a CP beam splitter.

Figure 4.18 studies the EM responses of both meta-atoms against the rotation angle θ. Figures 4.18a and 4.18b plot the amplitude and phase spectra of reflection and transmission for RCP and LCP excitations, respectively. It is observed clearly that both reflection phase ϕ_{RR}^r and the transmission phase ϕ_{RL}^t exhibit a relation of $\Delta\phi = 2\theta$ for meta-atom I, demonstrating that both the reflective RCP and transmissive LCP waves carry PB operations. The amplitude information of $|r_{RR}|$ and $|t_{LR}|$ is better than 0.92 at the frequency of f_0. Figures 4.18c and 4.18d illustrate the corresponding amplitude and phase spectra for meta-atom II under RCP and LCP incident waves, respectively. The amplitudes $|r_{RR}|$ and $|t_{LL}|$ keep at a high level and not affected by the rotation angle θ. The phase ϕ_{LL}^t is unchanged as θ increases, while ϕ_{RR}^r exhibits a variation shift of 2θ. The controllable phase combined with high amplitude makes the designed two meta-atoms very convenient to manipulate the CP waves in both transmission and reflection parts.

4.4.2 BIFUNCTIONAL META-LENS

We first design a bifunctional meta-lens that can converge the transmissive LCP wave and diverge the reflective RCP wave by using meta-atom I, as shown in Figs. 4.19a and 4.19b. To reach this goal, the transmission phase ϕ_{RL}^t and the reflection phase ϕ_{RR}^r should satisfy the following parabolic distributions

$$\begin{cases} \phi_{RL}^t(x, y) = k_0 \left(\sqrt{F_1^2 + x^2 + y^2} - F_1 \right) \\ \phi_{RR}^r(x, y) = -k_0 \left(\sqrt{F_1^2 + x^2 + y^2} - F_1 \right), \end{cases} \tag{4.8}$$

where $k_0 = w/c$ is the propagation constant and the focal length F_1 is set as 50 mm. Based on the working mechanism of PB operation, the rotation angle at the local position can be

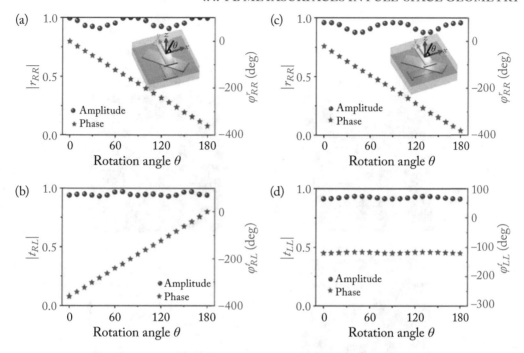

Figure 4.18: Amplitude and phase spectra of the proposed two meta-atoms as functions of rotation angles. (a, c) FDTD-simulated reflection amplitude $|r_{RR}|$ and phase ϕ^r_{RR} for RCP waves at 11 GHz for meta-atom I and meta-atom II, respectively. (b, d) FDTD-simulated transmission amplitude $|t_{RL}|/|t_{LL}|$ and transmission phase ϕ^t_{RL}/ϕ^t_{LL} for LCP waves at 11 GHz for meta-atom I and meta-atom II, respectively.

calculated as

$$\begin{cases} \theta^t_{RL}(x, y) = \frac{1}{2}\phi^t_{RL}(x, y) \\ \theta^r_{RR}(x, y) = \frac{1}{2}\phi^r_{RR}(x, y). \end{cases}$$

Then we discrete the metasurface as 14×14 meta-atoms and fabricated a metadevice sample with its picture shown in Fig. 4.19c. Our sample occupies a total volume of $154 \times 154 \times 4$ mm^3, corresponding to $2.65\lambda_0 \times 2.65\lambda_0 \times 0.15\lambda_0$. Figures 4.19e and 4.19f plot the phase distributions (ϕ^t_{RL} and ϕ^r_{RR}) at each meta-atom, which agree very well with theoretical values shown in Eq. (4.8). Quite different from the reported transmissive meta-lenses [24, 57], such as Huygens' meta-lens and gradient-index meta-lens, our design can largely suppress the transmission fluctuations since the amplitude is not affected by the phase profile.

With the sample in hand, we experimentally examine its EM functionality. We first consider its transmission functionality as a focusing lens. Shining our metadevice by a normally incident LCP wave through a horn antenna, we detect and record the local electric field ($\vec{E}$-

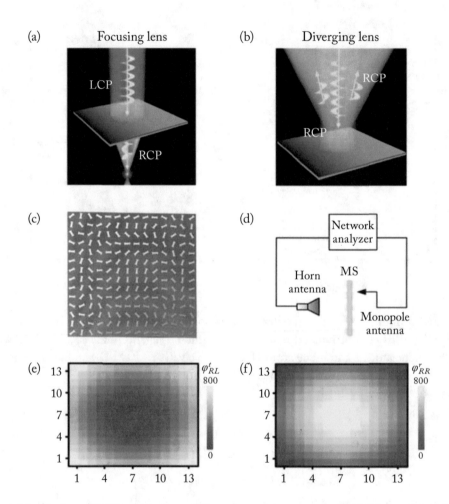

Figure 4.19: Design of bifunctional meta-lens by using meta-atom I. The proposed bifunctional meta-lens working as (a) a focusing lens at transmission part and (b) diverging lens at reflection mode for (a) LCP and (b) RCP incident waves, respectively. (c) Picture of the fabricated meta-lens sample. (d) Experimental setup for the near-field measurement. (e, f) Designed/fabricated transmission phase $\phi_{RL}^t(x, y)$ and the reflection phase $\phi_{RR}^r(x, y)$ at each meta-atom of our metadevice at the frequency of $f_0 = 11$ GHz.

field) by a 20-mm-long monopole antenna that is connected to a vector-field network analyzer (Agilent E8362C PNA), as shown in Fig. 4.20d. Figure 4.20a depicts the measured $Re(E_x)$ distributions at xoz plane at the frequency of $f_0 = 11$ GHz. We can see clearly that the incident LCP wave has been indeed converged to a focal point, which can be further demonstrated by the retrieved $|E_x|^2$ distributions at both xoz and yoz planes, as shown in Fig. 4.20c. The corresponding $Re(E_x)$ and $|E_x|^2$ simulations agree well with the measured results, as shown in Figs. 4.20b and 4.20d. To validate our design, we identify the focal length at the point with maximum $|E_x|^2$ at z-axis. Figures 4.20e and 4.20f plot the measured and simulated $|E_x|^2 \sim z$ curves. We find that the measured focal length (47 mm) agrees very well with the FDTD simulation (49 mm) and the theoretical case (50 mm). The slight difference is due to the finite-size of our metasurface and the imperfections of the incident waves. Based on the procedure shown in Section 3.3, we measure the absolute efficiency of our meta-lens as $\sim 88\%$ and the FDTD simulated value of about 91%. The missing power is taken away partly by the reflection and partly by the non-focused transmitted power ($\sim 6\%$ reflection for both simulation and measurement, 3% and 6% non-focused power for simulation and measurement).

We next consider the diverging performance of our meta-lens under an RCP incident wave. With a similar experimental setup as the transmissive focusing lens, we detect the $\vec{E}$-field at the reflection side of the metasurface. In order to see clearly the diverging effect, we purposely deducted the incident wave from the total field, ensuring a pure scattered field generated by our metasurface. As shown in Figs. 4.21a and 4.21b, good agreements between the measured and simulated $Re(E_x)$ distributions indicate the excellent diverging effects. We can see clearly that the normal scattering field has been almost completely suppressed due to the high-efficiency of our meta-atoms. Figures 4.21c and 4.21d plot the measured and simulated $|E_x|^2$ at xoz plane, which further demonstrates the excellently divergent effect of our metasurface. The efficiency, defined as the ratio between the power carried by reflected beam and incident beam, is measured and simulated as 94% and 96% at the frequency of 11 GHz. Compared with reported diverging lens [172], our design broke the 25%-efficiency limit and suppressed almost all of the undesired modes.

4.4.3   BIFUNCTIONAL CP BEAM SPLITTER

We now design a beam splitter to separate the CP waves with different polarizations to different sides of metasurface by using meta-atom II, as shown in Figs. 4.22a and 4.22b. Here, the beam deflection angle of the RCP wave can be freely manipulated. To design such a CP beam splitter, the transmission phase ϕ_{LL}^t and the reflection phase ϕ_{RR}^r should satisfy linear distributions

$$\begin{cases} \phi_{LL}^t(x, y) = C_1 \\ \phi_{RR}^r(x, y) = \xi x + C_2, \end{cases}$$

where C_1 and C_2 are constants and ξ is the phase gradient. ξ can be freely chosen and set as $\xi = 0.41k_0$, which can determine the deflection angle based on the generalized Snell's law

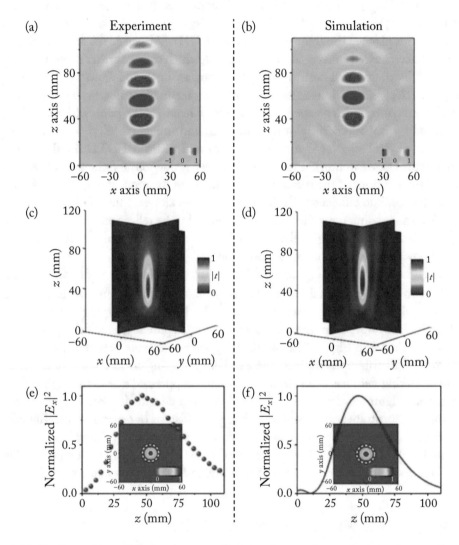

Figure 4.20: Performances of the transmissive focusing lens under excitation of normally incident LCP wave. (a) Measured and (b) FDTD simulated $Re(E_x)$ distributions on xoz plane when our metadevice is shined by an LCP incidence. (c) Measured and (d) FDTD simulated $|E_x|^2$ distributions on both xoz and yoz planes as our metadevice are shined by an LCP incidence. (e) Measured and (f) FDTD simulated $|E_x|^2$ at the focal point as a function of the focal position. Insets to (c) and (f) show the measured and FDTD simulated $|E_x|^2$ distributions at the focal plane of (c) $z = 47$ mm and (f) $z = 49$ mm, with the dashed-line circle denoting the focusing area.

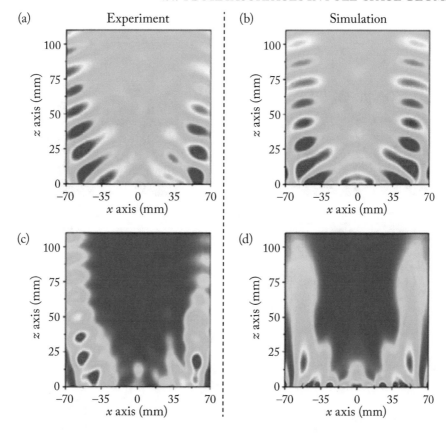

Figure 4.21: Performances of the reflective diverging lens under excitation of an RCP wave. (a, c) Measured and (b, d) FDTD simulated (a, b) $Re(E_x)$ and (c, d) $|E_x|^2$ distributions in xoz plane as our metasurface shined by normally incident RCP waves at $f_0 = 11$ GHz.

$\theta_r = \sin^{-1}(\xi/k_0)$ [21, 23, 24, 28, 30]. Our designed meta-atom II is very appropriate to achieve such a CP beam splitter. Based on the phase distributions, the supercell of our metasurface consists of six meta-atoms with each requiring a rotation angle of $\theta = 0.5 \times \phi$. Then we fabricate a microwave sample composing of 30×30 cells, occupying a total of $330 \times 330 \times 4$ mm^3, with its top view picture shown in Fig. 4.22c. As the transmission/reflection amplitude/phase shown in Figs. 4.22e and 4.22f, the fabricated phase distributions agree well with the theoretical values, while amplitudes ($|t_{LL}| > 0.93$, $|r_{RR}| > 0.95$) keep very high, indicating high-performance of our metadevice.

We first characterize the transmission performance of our sample under the excitation of an LCP wave. Shining our fabricated sample normally by an LCP wave through a CP horn antenna, we detect the angular power distributions by using another CP horn antenna within 360°

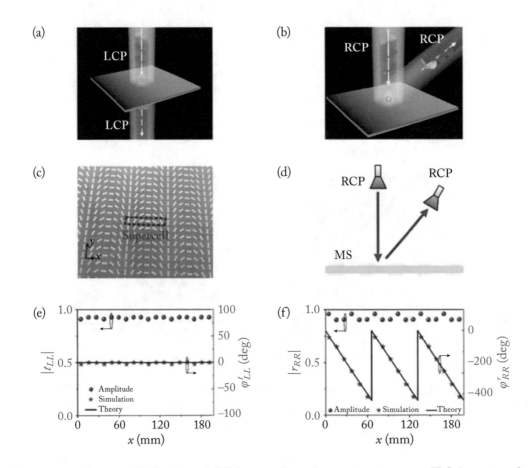

Figure 4.22: Design of bifunctional CP beam splitter by using meta-atom II. Schematic of our CP beam splitter, working as (a) a helicity keeper for LCP incidence and (b) a beam deflector for RCP incidence, respectively. (c) Picture of the fabricated sample of CP beam splitter. (d) Experimental setup to measure the scattering field. FDTD simulated (e) transmission amplitude $|t_{LL}(x)|$ and transmission phase $\phi_{LL}^{t}(x)$, and (f) reflection amplitude $|r_{RR}(x)|$ and reflection phase $\phi_{RR}^{r}(x)$ for the designed CP beam splitter, with the theoretical phase distributions calculated by $\phi_{LL}^{t}(x, y) = C_1$ and $\phi_{RR}^{r}(x, y) = \xi x + C_2$ (black lines).

region (with the experimental setup shown in Fig. 4.22d), with the measured results displayed in Figs. 4.23a and 4.23b. We note that almost all incident LCP waves have directly transmitted through our metasurface within a wide frequency interval of 10–12.4 GHz. Figures 4.23d and 4.23e depict the FDTD results, which shows an excellent agreement with the measured results. We quantitatively evaluate the working efficiency (shown in Fig. 4.23f) by the ratio between the power taken by the transmitted wave and the incident wave. The measured maximum efficiency reaches 90% at a frequency of 11 GHz, and the FDTD simulated efficiency is about 92%. The missing power is carried away partly by the reflection ($\sim$ 7% and 5% for measurement and simulation) and partly by cross-polarization conversion ($\sim$ 2% for both measurement and simulation). Referring to Fig. 4.23c, nearly flat wavefront at the transmission side is observed, demonstrating the desirable transmission mode for LCP wave excitation again.

We then characterize the deflection performance of our sample when shined by a normal RCP wave. The measured scattering fields (Figs. 4.23a and 4.23b) exhibit excellent agreements with FDTD simulations (Figs. 4.23d and 4.23e), indicating that all undesired scattering modes except for the anomalous reflection mode are totally suppressed at the frequency of 11 GHz. And the experimental deflection angle is consistent with the theoretical value (solid stars in Fig. 4.23) predicted by generalized Snell's law $\theta_r = \sin^{-1}(\xi_1/k_0)$. Using the same procedure discussed in Section 3.4, we retrieved the measured and simulated efficiency as shown in Fig. 4.23f. The maximum efficiency is better than 91% in the experiment and 93% in simulation at the working frequency of 11 GHz. Referring to Fig. 4.23c, we can see that the incident plane wave has been indeed reflected to an anomalous angle, and the pure deflection mode reinforces the high efficiency of our metadevice.

4.5   SUMMARY

To summarize, this chapter presented the general criteria to realize high-efficiency PB metasurfaces in reflection, transmission and full-space geometries, and the metadevices that we realized based on these criteria. Based on the desired meta-atoms working in different geometries, we experimentally realized several functional PB metadevices to efficiently control CP waves in different frequency regimes, achieving fascinating effects such as PSHE, vortex beam and Bessel beam generations, bifunctional meta-lens, and CP beam splitter.

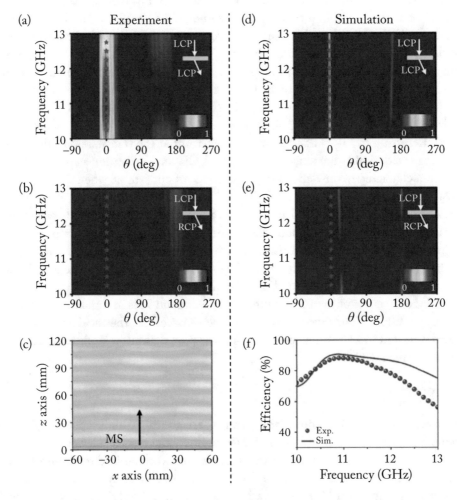

Figure 4.23: Performance of the helicity keeper at the transmission side for an LCP incident wave. (a, b) Measured scattered-field intensity obtained by (a) LCP and (b) RCP antennas (color map) at both sides of the metasurface shined by a normal LCP wave. (d, e) FDTD simulated scattered-field intensity obtained by (d) LCP and (e) RCP antennas (color map) at both sides of the metasurface under the excitation of a normal LCP wave. (c) The measured $Re(E_x)$ distributions at xoz plane at $f_0 = 11$ GHz when our sample is shined by a normal LCP wave. (f) The measured (blue circles) and FDTD simulated (red line) absolute efficiencies of our metasurface when shined by a normal LCP wave.

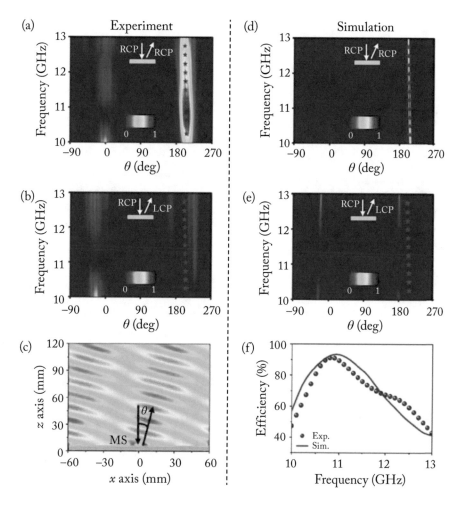

Figure 4.24: Performance of the beam deflector of our metasurface under excitation of an RCP wave. (a, b) Measured scattered-field intensity obtained by (a) RCP and (b) LCP antennas (color map) at both sides of the metasurface shined by a normal RCP wave. (d, e) FDTD simulated scattered-field intensity obtained by (d) RCP and (e) LCP modes (color map) at both sides of the metasurface under the excitation of a normal RCP wave. (c) The measured $Re(E_x)$ distributions at xoz plane at $f_0 = 11$ GHz at the reflection side of our metasurface when shined by a normal RCP wave. (f) The measured (blue circles) and FDTD simulated (red line) absolute efficiencies of our metasurface when shined by a normal RCP wave.

CHAPTER 5

Linearly Polarized Active Multifunctional Metasurfaces

The last two chapters summarized our efforts in realizing multifunctional metadevices, based on meta-atoms exhibiting distinct responses to EM waves with different polarizations or charity. However, these metasurfaces typically reply on *passive resonant* meta-atoms, whose intrinsic *dispersions* limit such *passive* metadevices' performances at frequencies other than the target one. Rigorously speaking, such design strategies only work for a *single* frequency, since the desired phase profile cannot maintain at other frequencies due to the intrinsic dispersions of passive resonant meta-atoms. To remedy these issues, one can construct *tunable* metadevices with functionality tuned by external knobs. Meanwhile, such tunable metadevices can also be viewed as multifunctional devices, although here the functionality switching is achieved via tuning external knobs rather than changing the polarization/charity of impinging EM wave. In this chapter, we introduce our efforts on utilizing the "active" scheme to achieve multi-functional and tunable metadevices with varactor diodes involved at microwave frequencies [148, 184]. In such a scheme, the dispersive response of each meta-atom is precisely controlled by an external voltage imparted on the diode. We start by discussing the inherent issues faced by passive meta-atoms in realizing metadevices and then present our solutions to overcome these issues as well as making tunable multi-functional metadevices. Our "tunable" approach, distinct from available strategies confined to homogenous metasurfaces and/or gradient-index materials with fabrication complexity and limited working frequencies [36, 85, 185–196], opens the door to achieve dynamical control on the dispersions and functionalities of inhomogeneous metasurfaces.

5.1 DESIGN PRINCIPLES: ROLE OF TUNABLE META-ATOMS ON PHASE COMPENSATIONS

Figure 5.1 schematically illustrates the underlying physics based on a reflective metasurface requiring a linear reflection-phase profile. Suppose a collection of meta-atoms have been chosen to form a device exhibiting an ideal linear-phase profile at the frequency f_0 (red line in Fig. 5.1a), such a *strictly* linear relationship cannot be satisfied at other frequencies (blue lines) with the same device since all resonant meta-atoms exhibit Lorenz-type phase dispersions (inset to Fig. 5.1a) and thus the phase difference between adjacent meta-atoms varies as a function of frequency. In particular, the phase gradient must decrease to zero at frequencies far away from f_0

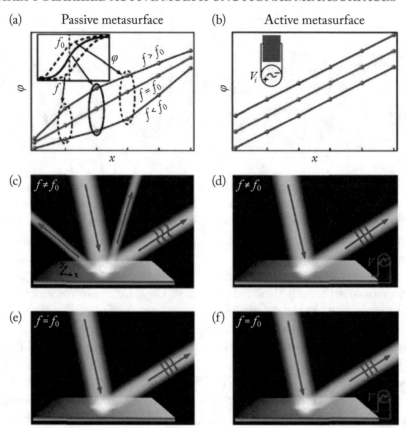

Figure 5.1: Role of the tunable scheme to overcome the phase distortions in passive metasurfaces. Typical phase responses of meta-atoms within a supercell in (a) a passive and (b) a tunable gradient metasurface, at the working frequency $f = f_0$ and non-working frequencies $f \neq f_0$. The nonlinear phase distributions at non-working frequencies in the passive device, caused by frequency dispersions of passive meta-atoms (see inset to Fig. 5.1a), can be rectified by active external controls in the tunable device. Typical scattering patterns of (c, e) passive or (d, f) tunable metasurfaces at (c, d) off-working frequencies and (e, f) working frequency.

where all resonant behaviors die off. As a result, while at f_0 the device can support single-mode anomalous reflection dictated by the generalized Snell's law (Fig. 5.1e), at other frequencies, the device's performance is significantly decreased with undesired modes appearing and deteriorated working efficiency (Fig. 5.1c). Such an issue seems inherent to all metasurfaces based on *passive* resonant meta-atoms which have been successfully implemented to design various metasurfaces working for different purposes [166, 167, 197–200], working in both reflection and transmission geometries and relying on either phase or amplitude modulations. Although the phase profiles in geometric-phase-based metasurfaces can immune from the frequency change, the real performances of such devices are dictated by the EM responses of their basic meta-atoms, which are still strongly frequency dependent. We emphasize that such "*chromatic aberrations*" cannot be solved by simply expanding the working bandwidths of metasurfaces using low-Q and/or multi-mode resonators since the frequency dispersion is intrinsic to all passive resonators.

The above limitation can be overcome by making *tunable* metasurfaces with meta-atoms controlled by external knobs (see inset to Fig. 5.1b). Borrowing similar *technical* ideas from previous studies on tunable reflectarrays/transmitarrays [201–205] to involve varactor diodes into the meta-atom structures, we can individually control the reflection phase of each active unit by applying an appropriate voltage on the diode, so that the distorted phase profiles of the passive metasurface at off-working frequencies ($f \neq f_0$) can be rectified (see Fig. 5.1b). Consequently, our tunable metasurface cannot only work well at the target frequency (Fig. 5.1f) but also works in the entire frequency band (Fig. 5.1d) with the single-mode operation and high efficiency.

To understand the dispersion issue in passive metasurfaces and lay a basis for designing our tunable structures, we first design a passive metasurface and analyze its wave-manipulation properties. Our meta-atom is a tri-layer structure, consisting of a composite planar resonator coupled with a metallic ground plane through a dielectric spacer (inset to Fig. 5.2a). To enlarge the working bandwidth of our device, we purposely design the meta-atom to contain two resonant modes, dictated by the metallic "H" structure and the metallic patch, respectively (Fig. 5.2a). The couplings between these planar structures and the ground plane generate two magnetic resonances, evidenced by the dips in the x-polarized reflectance spectra calculated by finite-difference-time-domain (FDTD) simulations, for two different systems with either only "H" or patch retained in the top layers (dashed and dotted lines in Fig. 5.2b). In FDTD calculations of reflection magnitudes/phases, we studied a single meta-atom with periodic conditions applied at its four boundaries and a floquet port assigned at a distance 15 mm away from the xy-plane where the meta-atom is placed. With both "H" and patch present, our meta-atom can possess a significantly enlarged working bandwidth ($\sim$ 2–7 GHz) due to the coupling of two resonances (solid line in Fig. 5.2b).However, we will show that such bandwidth enlargement *cannot* really improve the working performances of our metasurfaces, due to the intrinsic frequency dispersions of the passive resonant units.

Figures 5.3a and 5.3b depict how the reflection amplitude/phase of our passive meta-atom varies against parameter h_i and the target frequency. Obviously, $|r|$ is always near 1 which we do not need to worry about, but φ strongly depends on both h_i and frequency, since the resonant frequency of the meta-atom sensitively depends on the value of h_i. We now design a gradient reflective metasurface using the basic meta-atom structure. Setting the target frequency at $f_0 = 5.7$ GHz, we select 6 meta-atoms via carefully adjusting their structural details (see inset to Fig. 5.2a) based on the $\varphi \sim h_i$ relationships shown in Fig. 5.3b such that they exhibit the desired reflection phases following the relationship $\varphi(x) \sim \xi x$ with $\xi \approx 0.74k_0$ at this very

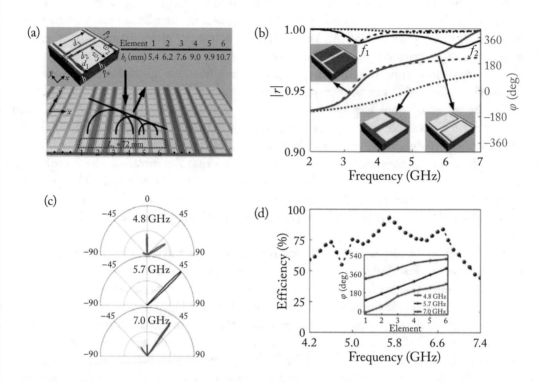

Figure 5.2: Dispersion-induced functionality deteriorations in a typical passive metasurface. (a) The geometry of the basic meta-atom and a gradient metasurface composed of six meta-atoms with parameter h_i varying from 5.4–10.7 mm. (b) FDTD-simulated spectra of reflection amplitude (left axis) and phase (right axis) of three periodic metasurfaces with unit-cells being the second meta-atom as shown in the inset to (a) (solid lines), meta-atom with patch only (dotted lines), and meta-atom with "H" only (dashed lines). (c) FDTD-simulated far-field scattering patterns of the gradient metasurface at three typical frequencies. (d) Calculated anomalous-reflection efficiency of the gradient metasurface as a function of frequency. The efficiency was defined as the ratio between anomalously reflected power ($\int_{(\theta_{i0}+\theta_{i1})/2}^{90°} P(\theta)d\theta$ with θ_{i0} and θ_{i1} being the reflection angles of normal and anomalous modes, respectively) and the totally reflected power ($\int_{-90°}^{90°} P(\theta)d\theta$), calculated by integrating the scattered-field intensity based on the FDTD-simulated patterns. Inset depicts the phase profiles at three selected frequencies. Geometrical parameters of the meta-atoms: $p_x = p_y = 12$, $w_1 = 0.8$, $w_2 = 0.5$, $w_3 = 5.1$, $d_1 = 0.25$, and $d_2 = 0.5$ mm. The spacer layer is assumed as the F4B dielectric board with $\varepsilon_r = 2.65$, $h = 6$ mm, and $\tan\delta = 0.001$, and all metallic films are assumed to be 36 μm thickness.

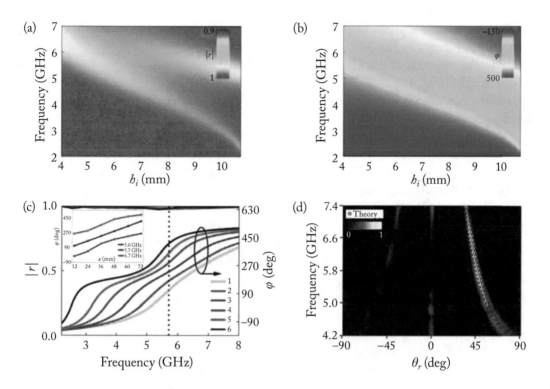

Figure 5.3: Additional FDTD results on the passive meta-atoms and metasurface when they are illuminated by a normally incident x-polarized wave. (a) Reflection amplitude ($|r|$) and (b) phase (φ) of a basic passive meta-atom vs. frequency and the parameter h_i. (c) Spectra of $|r|$ and φ for six passive meta-atoms, with inset showing the $\varphi(x)$ profiles of the passive metasurface at 5, 5.7, and 6.7 GHz. (d) 2D contour for the far-field reflected power vs. frequency and reflection angle θ_r for the passive metasurface. Blue stars are the results calculated with the generalized Snell's law. In the far-field calculations, we chose to study a line of meta-atoms belonging to five supercells arranged in the x-direction, with open boundary conditions set as its two ends and periodic boundary conditions set at its two boundaries along the y-direction. The scattered field intensity $P(\theta_r, \lambda)$ was normalized against P_0, which is obtained with the metasurface replaced by a PEC plate with the same dimensions.

frequency. Since the absorptions are generally weak due to low resonance quality factors of these structures (see Fig. 5.2b), we purposely adjusted the structures of six meta-atoms such that their reflection amplitudes are all nearly 1 (with fluctuations less than 0.1). The amplitude fluctuations over meta-atoms thus have a much less pronounced effect on the device's working efficiency than the corresponding phase variations. The six meta-atoms form a supercell of our passive metasurface, as shown in Fig. 5.2a. To justify our design, we plotted in Fig. 5.3c the reflection amplitude/phase spectra of our six meta-atoms, from which we find a perfectly linear phase gradient of $\pi/3$ (109.3°, 169.2°, 229.4°, 289.6°, 349.5°, and 409.6°) at 5.7 GHz, with all reflection amplitudes larger than 0.97. However, such a linear phase gradient cannot be strictly held at other frequencies.

We then employed FDTD simulations to study the scattering patterns of such metasurface illuminated by x-polarized EM waves, at different frequencies. While the scattering pattern at $f_0 = 5.7$ GHz (middle panel in Fig. 5.2c) does exhibit a single-mode reflection at the designed angle ($\theta_r \approx 47°$), multi-mode diffractions are significant at other frequencies (upper and lower panels in Fig. 5.2c). The intuitioniztic support can be found from the 2D contour of scattering patterns shown in Fig. 5.3d, where a high anomalous reflection of $m = +1$ mode with well-suppressed diffractions of $m = 0$ and $m = -1$ mode only around 5.7 GHz, where a linear phase gradient is strictly held. At other frequencies, diffractions to the "wrong" channels are significant which decreases the desired working efficiency. As a result, the true working efficiency of our device drops dramatically as frequency leaves 5.7 GHz, as shown in Fig. 5.2d. The physics can be understood from Fig. 5.3c. Although a perfect linear profile $\varphi(x)$ is found at 5.7 GHz, it is obviously distorted at 4.8 GHz and 7 GHz, due to the intrinsic dispersions of the meta-atoms. In particular, the phase gradient decreases to zero at the boundaries of the working band where all resonances die off. Such phase distortions are responsible for the appearances of diffractions to the *wrong* channels, which in turn, decreases the anomalous-mode conversion efficiency from 93.5% to 54.8% and 62.9%, respectively. The two additional side peaks at 5 and 6.7 GHz in Fig. 5.2d is caused by the enhanced linearity (see inset to Fig. 5.3c), accidentally existing for such dual-mode meta-atoms.

Our tunable strategy to overcome the dispersion-induced phase distortions can be seen in Fig. 5.5a, where the tunable meta-atoms are topologically equivalent to the passive one, only with the central bars of the "H" structures broken and then connected by varactor diodes (SMV1430-079LF, Skyworks Solutions Inc. [206]), which are, in turn, *independently* controlled by six external voltages $V_i (i = 1, \ldots, 6)$. Since the SMV1430-079LF only works well below 10 GHz, it will induce unreliable LC values and large self-resonant loss above, which is undesirable for our applications. The higher frequency micro-electromechanical system diode affords an extremely low loss strategy and thus would benefit much the high absolute efficiency. To offer a guideline to control the two resonant modes in our tunable meta-atoms for the metasurface design and phase compensations, here we establish the equivalent CM for our structure (Fig. 5.4a). The resonant behavior and tunability can be understood from a lumped LC circuit because our

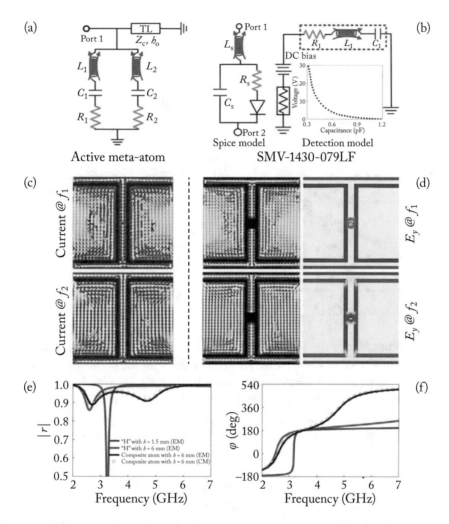

Figure 5.4: Analyzing the two resonant modes with FDTD and circuit model studies. The equivalent CM for (a) active/passive meta-atom and (b) varactor SMV1430-079LF. The inset shows the C-V curve of the varactor. Field/current distributions for the (c) passive and (d) active meta-atoms, respectively. CM and EM simulated reflection (e) amplitude ($|r|$) and (f) phase (φ) of the active meta-atom and the "H"-shaped meta-atom with different h. Other geometrical parameters are $p_x = p_y = 12$ mm, $w_1 = 0.8$ mm, $w_2 = 0.5$ mm, $w_3 = 5.1$ mm, $d_1 = 0.25$ mm, $d_2 = 0.5$ mm, $d_3 = 10$ mm, and $h_i = 10.5$ mm. The circuit parameters are retrieved as $L_1 = 18.76$ nH, $C_1 = 0.111$ pF, $L_2 = 0.059$ nH, $C_2 = 0.196$ pF, $R_1 = 8.37$ Ω, $R_2 = 0.114$ Ω, $Z_c = 204.9$ Ω, and $h_o = 58.9°$.

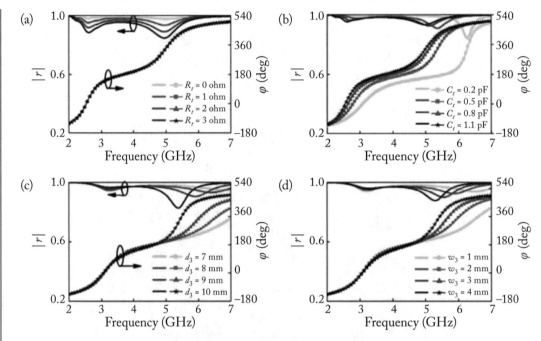

Figure 5.5: EM properties of the active meta-atom depending on different parameters. FDTD calculated spectra of reflection amplitude $|r|$ and phase φ for an active meta-atom with different (a) R_t ($d_3 = 10.16$ mm, $w_3 = 5.1$ mm, and $C_t = 1.2$ pF fixed), (b) C_t ($d_3 = 10.16$ mm, $w_3 = 5.1$ mm, and $R_t = 1$ Ω fixed), (c) d_3 ($w_3 = 5.1$ mm, $R_t = 1$ Ω, and $C_t = 0.31$ pF fixed), and (d) w_3 ($d_3 = 10$ mm, $R_t = 1$ Ω, and $C_t = 0.31$ pF fixed). Other geometrical parameters are $p_x = p_y = 12$ mm, $w_1 = 0.8$ mm, $w_2 = 0.5$ mm, $d_1 = 0.25$ mm, and $d_2 = 0.5$ mm.

meta-atom is much smaller than the wavelength of interest. The two magnetic resonances (at f_1 and f_2) can be modeled by two series resonant tanks formed by L_1, C_1, and R_1, and L_2, C_2, and R_2, respectively. The back metallic layer was represented by the ground, whereas the transmission through the dielectric substrate (with impedance Z_o and thickness h) was modeled by a transmission line (TL) with equivalent impedance Z_c and electrical length h_o.

In Fig. 5.4b, R_t, L_t, and C_t in the detection model represent the total resistance, total inductance, and total capacitance of the whole system, which have included the contributions from the junction resistance R_s, the lead inductance L_s and the package capacitance C_s in the spice model. In our design, C_s is negligible compared to the junction capacitance C_j of the varactor SMV1430-079LF, yielding $C_t \approx C_j$. In full-wave FDTD simulations, the varactor is replaced by a series resonant tank with $L_s = L_t = 0.7$ nH, $R_s = R_t = 1.5$ Ω, and 0.3 pF $\leq C_t \leq 1.2$ pF depending on the externally applied voltage. Using Agilent's Advanced Design System (ADS), we obtained the $C_t \sim V$ curve for our varactor (inset to Fig. 5.4b), which is extremely helpful for our design. As is shown, C_t decreases from 1.2 to 0.3 pF when V increases from 0 to 30 V. Moreover, C_t changes sharply when V is small and saturates at a certain value when V approaches 15 V. Therefore, one can choose a small V to obtain a sharp tunability. The electric-current distributions associated with the same resonant mode remain the same in the passive (Fig. 5.4c) and active (Fig. 5.4d) meta-atoms, although the resonant frequencies are at different positions. The induced currents are mainly around the "H" in the first mode (f_1) but are mainly on the patch surface in the second one (f_2), consistent with the picture set up in Fig. 5.2b. However, we found strong couplings between the patch and the "H" at the second mode (f_2), evidenced by strong overlapping of electric fields belonging to two different structures. Such coupling blue-shifted f_2 in the composite resonator relative to the patch-only case. However, f_1 was hardly affected by the existence of the patch since such coupling is remarkably weak at f_1. Based on Figs. 5.4c and 5.4d, we got $L_1 \approx L_w + L_t$, $L_2 \approx L_p/2$, $C_1 = C_f * C_t/(C_f + C_t)$, and $C_2 \approx 2(C_p + C_c)$, where L_w is contributed by the perpendicular bar, C_f models the fringing capacitor formed between parallel horizontal bars of two adjacent elements, L_p and C_p are the inductor and capacitor of the patch, respectively. The losses of the above two tanks are modeled by R_1 and R_2, respectively. TL theory tells us that $f_1 = 1/2\pi\sqrt{L_1 C_1}$ and $f_2 = 1/2\pi\sqrt{L_2 C_2}$, and thus the response of our active meta-atom (at an arbitrary frequency) can be efficiently tuned by engineering these parameters (i.e., L_1, C_1, L_2, C_2 etc.).

Figures 5.4e and 5.4f show the spectra of reflection amplitude and phase of a typical *active* meta-atom with $h = 6$ mm and biasing voltage of $V_i = 0$ V, obtained by FDTD simulations (black lines) and CM analyses (open circles). Reasonable agreement between the two results validated our CM. Compared to its passive counterpart (see Fig. 5.2), our active meta-atom exhibits similar EM responses but with an enhanced Q factor due to the additional LC elements contributed by the active diode. Therefore, cascading two resonances in our design is crucial to enlarge the phase/frequency tuning range. In addition, increasing the spacer thickness (h) is another approach to decrease the Q factor.

To understand the roles of different parameters to control the EM responses of our active meta-atoms, we performed extensive simulations by varying the circuit parameters R_t, C_t and the patch parameters w_3, d_3 while maintaining other parameters unchanged. Figure 5.5a shows that increasing R_t can increase the absorption dip but has nearly no effect on the phase spectra. Meanwhile, Fig. 5.5b shows that both f_1 and f_2 undergo a red-shift when C_t increases from 0.2–1.1 pF. The capacitive coupling between the patch and the "H" enables such a simultaneous control. Most importantly, the reflection phase can be progressively tuned to cover a 270° phase modulation within 2–7 GHz. Finally, Figs. 5.5c and 5.5d show that varying w_3 or d_3 can only modulate the resonance at f_2 or f_1, consistent with the conclusion drawn in Fig. 5.2. To sum up, the external biasing (Fig. 5.5b) provides a further way to control the reflection phase of our meta-atom in addition to the geometrical tuning (Figs. 5.2c and 5.2d), and the possibility to separately control two different resonances helps us achieve arbitrary values for f_1/f_2.

5.2 A BIFUNCTIONAL METADEVICE EXHIBITING ANOMALOUS-REFLECTION AND SURFACE-WAVE CONVERSION FUNCTIONALITIES

Based on the above role of tunable meta-atoms for phase compensation, we now experimentally demonstrate that the functionality of our device can be dynamically switched from a conventional reflector to a surface wave (SW) coupler, again based on the tunable and precise local-phase control. Such switchable functionality, not realized in previous SW couplers, can be very useful in stealth applications where both normal detection and radar cross-section reduction are required.

Our bifunctional metadevice contains a periodic array of supercell consisting of tunable "H" meta-atoms which are, in turn, *independently* controlled by six external voltages $V_i (i = 1, \ldots, 6)$, see Fig. 5.6a. Since the lumped-capacitance C_t of the diode depends sensitively on the voltage applied through it, the resonance frequencies of our meta-atoms can be controlled by these external voltages. As a result, the reflection-phase spectra of six meta-atoms can be efficiently and *independently* controlled by V_i. Figure 5.6b depicts how the phase of the second meta-atom $\varphi_2(f)$ varies against V_2. As V_2 increases from 0 to 30 V, the lumped capacitance of the diode decreases from 1.2 to 0.3 pF, which shifts the relevant resonance from 4.7 to 6.6 GHz and thus generates dramatic modulations on $\varphi_2(f)$. Such modulation strongly depends on the structural details of the meta-atom.

To yield the desired phase distributions $\varphi(x)$ of the tunable metasurfaces, we established a CAD (computer-aided design) method to search for the required biasing voltages $\{V_i\}$ imparted on the meta-atoms. The design procedure consists of the following four steps (Fig. 5.7). First, we determine the periodicity p_i of meta-atoms, phase gradient φ_0 (the number of elements N used in a supercell) at initial frequency f_0, and capacitance C_0. To engineer the linear/parabolic phase gradient at f_0, we can readily synthesize the initial geometrical parameters of above N elements with the help of CM theory by performing FDTD optimizations. In this particular design, C_0 is chosen as the maximum value that a varactor can afford such that the phase gradient can be restored at $f > f_0$. Second, we obtain the phase responses of the above N elements vs. frequency and capacitance C_t by FDTD parametric analyses. Then, a capacitance-phase (C_t–φ) relation curve is obtained. Third, we achieve all possible φ/C_t solutions of N elements at each frequency according to the C_t–φ curve by employing the spline interpolating method and traversal query (a root-finding algorithm). Since the φ/C_t solutions of N elements satisfying the perfect phase gradient may not be unique at each frequency, we select the solution that exhibits minimum C_t span between N atoms to guarantee the optimum bandwidth. The root-finding process will be continued at the next frequency if the optimum C_t of N elements all falls in the capacitance tuning range (0.3 pF $< C_t <$ 1.2 pF in this particular design) at the present frequency. Otherwise, it will be terminated and the operation bandwidth of the active metasurface is determined. In the fourth step, we derive the required voltages of N elements at each frequency according to the capacitance-voltage (C-V) curve of the varactor. To guarantee

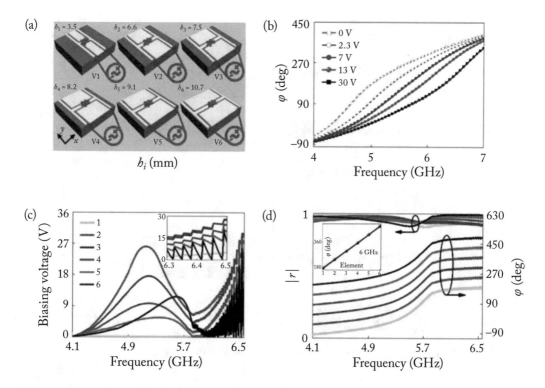

Figure 5.6: Designs and tunability of tunable meta-atoms. (a) Geometries of six tunable meta-atoms, each with a varactor diode loaded to connect the broken central line of the "H". (b) Calculated reflection-phase spectra of a metasurface consisting of a periodic array of the second meta-atoms at different biasing voltages imparted on the diodes. (c) Retrieved values of the requested voltages imparted on six meta-atoms to maintain the linear phase distributions at every frequency within the band 4.1–6.6 GHz. Inset shows a zoom-in view of the retrieved voltages at frequencies from 6.3–6.5 GHz. (d) FDTD calculated reflection amplitudes and phases of six meta-atoms biased by the corresponding voltages as shown in (c). Inset plots the phase profile of the tunable metasurface at 6 GHz.

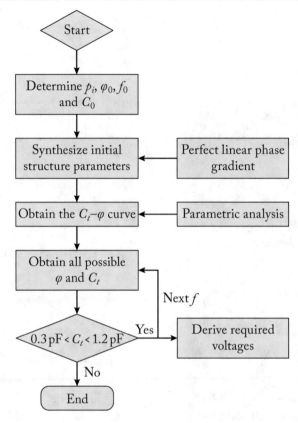

Figure 5.7: The CAD flow chart for broadband phase gradient design.

sufficient precision, numerical interpolation is also adopted for the C-V relation with a mass of samples.

Following the above active CAD design strategy, we first vary h_i of six meta-atoms (keeping other parameters unchanged) to ensure that the phase profile of our metasurface is perfectly linear at an initial frequency $f_0 = 4.1$ GHz under $V_i = 0$ V voltage ($C_t \approx 1.2$ pF). Figures 5.8a and 5.8b depict how $|r|$ and φ of the "active" meta-atom (with $V_i = 0$ V fixed) vary against its h_i and frequency while keeping other geometrical parameters unchanged. We note that the "active" meta-atom, when it is under fixed biasing voltage, behaves similarly to a passive one, except for slightly enhanced absorption brought by the inserted active element. Figures 5.8a and 5.8b assist us to fix the geometries of our six meta-atoms (denoted by 1, 2, 3, 4, 5, and 6 for convenience), which exhibit $h_i = 3.5, 6.6, 7.5, 8.2, 9.1,$ and 10.7 mm, respectively (all other parameters are fixed). To validate our design, we show in Fig. 5.8c the calculated spectra of reflection amplitudes and phases for the six active meta-atoms with zero biasing voltages. At $f_0 = 4.1$ GHz, we find that the phases of these meta-atoms do exhibit a perfectly linear phase

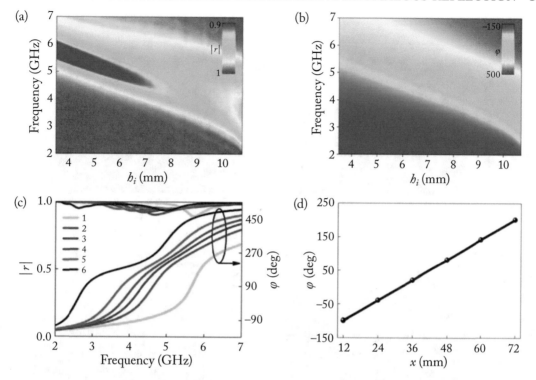

Figure 5.8: FDTD results of the active meta-atoms studied in the main text under zero biasing voltages. Computed reflection (a) amplitude ($|r|$) and (b) phase (φ) of an active meta-atom (under zero biasing voltages) vs. frequency and its parameter h_i. The supercell of the metasurface is formed by six meta-atoms with $h_i = 3.5, 6.6, 7.5, 8.2, 9.1$, and 10.7 mm, respectively. (c) FDTD simulated spectra of $|r|$ and φ of six meta-atoms under zero biasing voltages. (d) Phase profile $\varphi(x)$ of the active metasurface formed by the six meta-atoms, under zero biasing voltages at the frequency 4.1 GHz.

gradient (see Fig. 5.8d), while their reflection amplitudes are all larger than 0.9 with fluctuations less than 0.1. We next search for the voltage combinations $\{V_i(f)\}$ to yield perfect linear phase relationships at frequencies $f > f_0$. Figures 5.9a–f depict how the reflection phases of six meta-atoms vary against C_t and frequency. As expected, tuning each C_t (through varying the voltage V imparted on the very meta-atom) can significantly modulate the reflection phase of this active meta-atom within the frequency range 3.2–7 GHz.

Based on the $\{\varphi_i(f) \sim C_t\}$ in Figs. 5.8 and 5.9, and the $C_t \sim V$ relation shown in Fig. 5.4b, we finally obtained $\{\varphi_i(f) \sim V_i\}$ relationships for all six meta-atoms. With all $\{\varphi_i(f) \sim V_i\}$ relationships known, we can retrieve the voltage combinations $\{V_i\}$ imparted on six meta-atoms at every frequency, which can yield a *perfectly* linear phase profile for the meta-

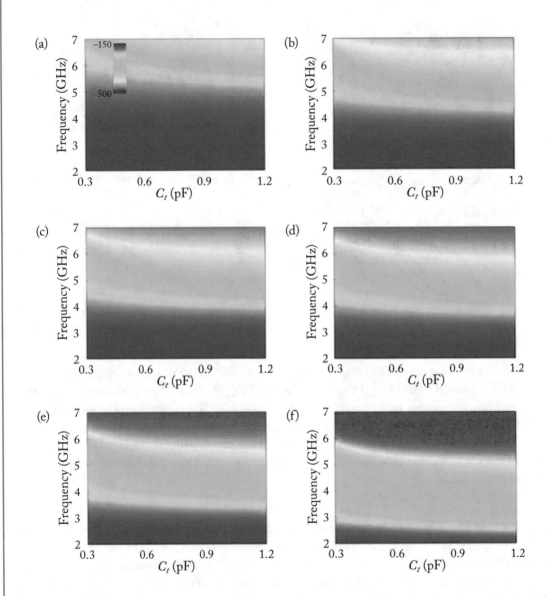

Figure 5.9: FDTD parametric analyses on six active meta-atoms. The reflection phases vs. C_t and frequency for six active meta-atoms with (a) $h_i = 3.5$ mm, (b) $h_i = 6.6$ mm, (c) $h_i = 7.5$ mm, (d) $h_i = 8.2$ mm, (e) $h_i = 9.1$ mm, and (f) $h_i = 10.7$ mm.

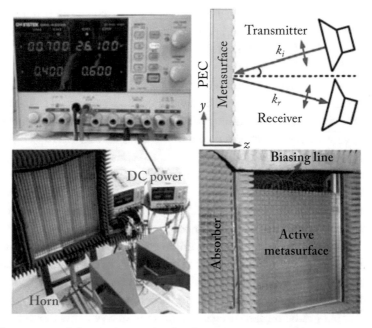

Figure 5.10: Illustration of the angle-resolved reflection measurement setup.

surface constructed by these elements. We note that the solutions $\{V_i\}$ may not be unique since what we required is mere that the phase *differences* between adjacent elements are a constant (i.e., $\varphi_{i+1} - \varphi_i = \pi/3$, $i = 1, \ldots, 5$). Figure 5.6c depicts the retrieved $\{V_i\}$ varying against frequency, which is one particular set of solutions that we found by restricting on moderate values of voltages. We note that large fluctuations exist in V_6 at high frequencies. This is because the resonance frequency of the sixth meta-atom is too low, and thus the meta-atom becomes quite *inactive* with respect to external control at high frequencies. To validate our solution, we plot in Fig. 5.6d the spectra of reflection amplitudes and phases for six meta-atoms controlled by the external voltages given in Fig. 5.6c. Perfect linear phase gradient is found for every frequency within the band 4.1–6.6 GHz (see inset for the phase profile at 6 GHz), with nearly uniform reflection amplitudes ($|r| > 0.9$). This is quite unusual since the original resonances of these meta-atoms are completely different. Yet, under appropriate biasing voltages, we can make their phase differences keeping constant within a very broadband, overcoming the intrinsic *chromatic aberrations* in passive metasurfaces. In principle, one can use our strategy to make a meta-atom to exhibit an arbitrary reflection phase φ, and thus the working bandwidth can be made as large as possible. However, this is practically impossible since any diode exhibits a limited tunable range and also the external voltage cannot be made too large. These factors collectively set an upper limit on the working bandwidth of a tunable metasurface. However, we emphasize here that within such a working bandwidth the tunable metasurface exhibits an *ideal* performance without *chromatic aberrations*. Therefore, our bandwidth is intrinsically different from conventional working bandwidth for a wideband passive device in which the device's performance gets deteriorated at frequencies deviating from the central target one.

We fabricated a tunable metasurface according to the design (Fig. 5.6a) and experimentally characterized its wave-manipulation performances. The active metasurfaces were fabricated using a printed-circuit-board technique with all varactor diodes attached to the top metallic microstructure using surface-mount technology. To guarantee perfect electrical connections with correct electrodes, all varactors are checked by a multimeter of VICTOR VC9807A+. To control the capacitances of the varactor diodes, we connected constant-voltage sources (GPD4303S, GWINSTEK company) to their two ends by two thin wires. To suppress the leakage of microwave signals and thus enable a robust and reliable performance, the top and bottom horizontal bars functioning as zero- and reverse-biased lines, respectively, are engineered thinner than the perpendicular bar to provide a high reactance.

As shown in Fig. 5.11a, the fabricated sample contains 30×30 meta-atoms, with a total size of 360×360 mm^2. Since the phase gradient is along the x-direction, the meta-atoms in each vertical column are identical. Considering further the super-periodicity, we only need six independent constant-voltage sources to bias those diodes belonging to the same type of meta-atoms. We understand that our gradient metasurface functions as an SW converter at frequencies below $f_c = 4.167$ GHz when the phase gradient is larger than the free-space wave-vector, while it is an anomalous reflector at frequencies above f_c. We first focus on the frequency region above f_c in this section. Biasing the meta-atoms at their corresponding voltages as shown in Fig. 5.6c, we performed both microwave experiments and FDTD simulations to study the scattering patterns of our metasurface at different frequencies (see Methods section for microwave characterization details).

In the far-field experiments, see the measurement setup shown in Fig. 5.10, shining the metasurface by a microwave beam (with spot size ~ 250 mm at 6 GHz (E plane)) emitted from a linear-polarization horn antenna placed 1.2 m away, we measured the scattered-field intensity with another identical horn antenna, which can be freely rotated (with a step of 3°) on a circular track with 1.2 m radius. All scattered power was recorded by a vector network analyzer (Agilent E8362C PNA) and was normalized against a reference signal P_0, obtained by measuring the specularly reflected signal with the metasurface replaced by a metallic plate with the same size. In measuring P_0, two horn antennas were placed on the same side of the sample with a 10° angle separation. In the near-field experiments, we placed a 15-mm-long monopole antenna perpendicular to the metasurface, moving at a plane 3 mm above the metasurface, to measure the local E_z field (with both amplitude and phase). Absorbing materials are placed at the right side of the mushroom structure to prevent any reflected signals.

Figure 5.11b depicts the measured scattering patterns of the tunable metasurface under normal-incidence excitations at 5 frequencies (4.5, 5.0, 5.5, 6.0, 6.5 GHz). In all the cases studied, we found that the reflection beams only contain the desired anomalous mode with diffrac-

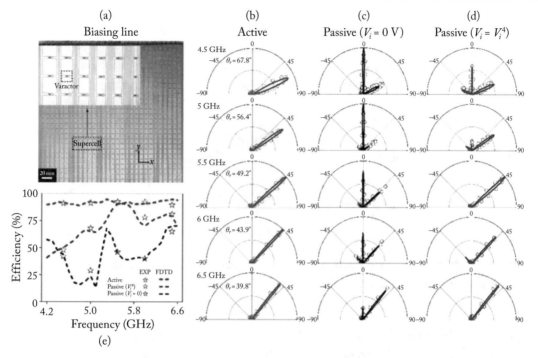

Figure 5.11: Far-field characterizations on the tunable metasurface. (a) Photograph of the fabricated tunable metasurface, with inset depicting a supercell of the sample. Measured (symbols) and FDTD-simulated (line) scattering patterns of the tunable metasurface at 5 frequencies, when the sample is biased (b) at the required voltage combination corresponding to each frequency, (c) at 0 V voltage, and (d) at the fixed voltage combination (denoted by $\{V_i^4\}$) corresponding to the frequency of 5.5 GHz. (e) Frequency-dependent anomalous-reflection efficiency of our device for the cases studied in (b) (red), (c) (blue), and (d) (black), obtained by analyzing the measured (symbols) and FDTD simulated (lines) scattering patterns.

tions to the *wrong* channels fully suppressed. The experimentally identified anomalous reflection angle θ_r changes from 67.8 to 39.9° as frequency f increases from 4.5 to 6.5 GHz, agreeing well with the generalized Snell's law: $k_0 \sin \theta_r = \xi$ with $k_0 = 2\pi f/c$ being the wave vector in free space and c the speed of light (see Fig. 5.12). Slight deviations at low frequencies are induced by the finite-size effect since the size of our metasurface becomes *effectively* shortened in terms of wavelength at low frequencies. Since the phase gradient, $\xi = 2\pi/L$ (L is the super periodicity) of our device is frequency independent, θ_r must be a decreasing function of f, which has been verified experimentally. We note that the anomalous-reflection beam becomes narrowed as f increases (Fig. 5.11b). This is again quite physical since our metasurface becomes *effectively* en-

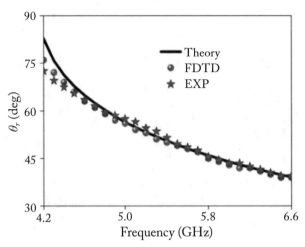

Figure 5.12: Verifications on generalized Snell's Law. $\theta_r \sim f$ relations obtained by experiments, FDTD, and the generalized Snell's law for the active metasurface.

larged in size (in terms of wavelength) at higher frequencies, although its physical size remains unchanged. FDTD simulations match very well with the experimental data.

Figure 5.11b already indicates that our tunable metasurface exhibits nearly ideal functionalities, much more superior than its passive counterpart (see Fig. 5.2). Obviously, such improvement comes from the perfect linear phase gradient enabled by external controls. To further highlight the importance of the tunable phase control, we repeated all characterizations on the same device which is now under zero-voltage biasing. It is not surprising to see that the device does not show single-mode anomalous reflections at those frequencies selected (Fig. 5.11c). Instead, diffractions to the wrong channels are significant, similar to the passive device. We note that zero-voltage-biased metasurface is not an ideal candidate to mimic a passive metasurface since at this biasing condition the metasurface does not exhibit a good linear phase distribution at *any* frequency. Therefore, we further study the scattering properties of the same device under the biasing voltages designed for 5.5 GHz. Such a device can better mimic a passive system since the effective lumped parameters of the diodes do not change with frequency. Figure 5.11d shows the measured/simulated reflection spectra of the device at the same five frequencies. As expected, the device works with ideal functionality only at the frequency of 5.5 GHz, but the performance gets deteriorated significantly at other frequencies.

Such comparisons highlighted the key advantage of our tunable scheme—our device can work with the best functionality at any frequency within broadband. Figure 5.11e compares the working efficiencies of three devices studied in Figs. 5.11b–d as functions of frequency. Whereas our tunable device (under appropriate biasing voltages) can always exhibit high anomalous-reflection efficiency ($\sim 90\%$) within 4.2–6.6 GHz, the fixed-voltage-biased device only works for a single frequency similar to a passive device, and the unbiased device exhibits poor efficiencies for all frequencies within the band.

We now focus on another frequency domain ($f < f_c$) and demonstrate a further important application of our tunable scheme, that is, dynamical functionality switching. The working principle can be understood as follows. At a frequency $f < f_c$, our gradient metasurface can function as a propagating wave (PW) to the SW converter when its phase profile $\varphi(x)$ exhibits a perfect linear relationship but becomes a normal reflector when φ remains nearly a constant within a supercell. Therefore, engineering $\varphi(x)$ in a designed manner can create two "states" between which the metasurface can switch.

To experimentally demonstrate this idea at 4.1 GHz, we first design and fabricate a mushroom structure (see inset for its unit-cell geometry) supporting spoof surface plasmon polariton (SPP) propagations at this frequency, and then place it at the right-hand side of the gradient metasurface (Fig. 5.13a). Such a device can guide out the SW generated on the metasurface when the latter is shined by a normally incident PW. The real experimental sample and setup are shown in Fig. 5.13b. In our experiment, shining the metasurface *alone* by an x-polarized PW, we employed a monopole antenna to measure the distribution of local $Re(E_z)$ fields generated on both the metasurface and the mushroom surface. To ensure a high conversion efficiency, the mushroom structure is carefully designed such that the wave vector of its spoof SPP is $k_{spp} = 1.02k_0$ at 4.1 GHz (see Fig. 5.14), which is well matched to the wavevector of the "driven" SW generated on the metasurface (the phase gradient ξ of the active metasurface).

We performed near-field (NF) measurements in two cases where the meta-atoms are biased at two different voltage combinations. In the "On"-state, the biasing voltages are chosen from Fig. 5.13c so that $\varphi(x)$ exhibits a perfect linear relationship (inset to Fig. 5.13c). In the "Off"-state where all voltages are set as 30 V, however, the $\varphi(x)$ relation deviates significantly from a linear one and φ remains nearly unchanged in the large area of a supercell (inset to Fig. 5.13d). As expected, when the metasurface is set at the "On"-state, we do observe very strong spoof SPP signals and the simulated/measured NF patterns (Figs. 5.13c and 5.13e) exhibit a very well-defined $k_{spp} = 1.02k_0$, indicating that our device is a very efficient SW converter. On the other hand, when the metasurface is switched to the "Off"-state, we only detected negligible SPP signals on the mushroom structure with an On/Off power ratio 22.1, implying that the device now ceases to behave as an SW converter (Figs. 5.13d and 5.13f). To further differentiate the functionalities of our device at two states, we performed far-field (FF) experiments to measure the scattering patterns (Figs. 5.13g and 5.13h) of our device at two different states. Combining experimental data from NF and FF measurements, we conclude that our device is mostly an SW converter in the "On"-state but dynamically changes to a conventional reflector in the "Off"-state.

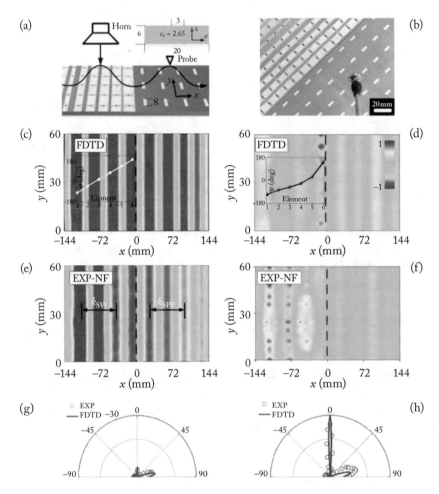

Figure 5.13: **Near-field characterizations on the tunable metasurface: Dynamical functionality switching.** (a) Schematics of the PW-SW conversion setup. Inset shows a unit cell of the mushroom structure, consisting of a metal bar (sized 3 mm × 8 mm) coupled with a continuous metal plate through a 6 mm—thick dielectric spacer ($\varepsilon_r = 2.65$). (b) Picture of the fabricated sample and the near-field probe. (c, d) FDTD simulated, and (e, f) measured $Re(E_z)$ distributions under (c, e) the voltage combination corresponding to frequency 4.1 GHz, as shown in Fig. 5.6c and (d, f) $V_i = 30$ V for all elements. Insets in (c) and (d) depict the phase profiles under the specified external voltages. All fields in (c–f) share the same color bar shown in (d). FDTD simulated and measured far-field scattering patterns of the metasurface at 4.1 GHz under (g) the voltage combination corresponding to frequency 4.1 GHz and (h) $V_i = 30$ V for all elements. In near-field FDTD calculations, the studied system contains five/three supercells along the y/x direction, again with periodic/open boundary conditions imposed at the boundaries.

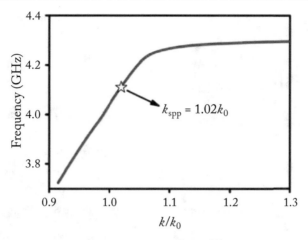

Figure 5.14: FDTD-computed dispersion curve of spoof SPP on the mushroom structure adopted in the inset to Fig. 5.13a. At 4.1 GHz, we do find $k_{spp} = 1.02k_0 = \xi$.

5.3   ABERRATION-FREE AND FUNCTIONALITY-SWITCHABLE MICROWAVE META-LENSES

In this section, we further utilize the active scheme established in Section 5.1 to overcome such dispersion-induced issues. By making tunable metasurfaces with all individual meta-atoms independently controlled by external knobs, we can precisely control the phase profiles of the metasurfaces at different frequencies, so that not only the dispersion-induced phase distortions can be rectified but also the functionalities of the metadevices can be dynamically switched. The dual-mode reflective-type meta-atom with geometrical details shown in Fig. 5.15a is very beneficial for the relatively large frequency/phase tuning range and nearly uniform reflection magnitudes, which is the key for aberration-free and functionality-switchable meta-lenses. Again, the building block is a tri-layer meta-atom consisting of a composite planar resonator (containing a metallic "H" structure and a metallic patch) coupled with a metallic ground plane through a dielectric spacer ($\varepsilon_r = 2.65$, $h = 6$ mm, loss tangent $\tan \delta = 0.001$). To make the meta-atom tunable, we break the central bar of the "H" structure and then connect its two parts by a varactor diode (SMV1430-079LF, Skyworks Solutions Inc.), which is then controlled by an external voltage source. The couplings between the top structure and the ground plane generate two magnetic resonances at frequencies f_1 and f_2, evidenced by the dips in the FDTD simulated x-polarized reflection spectrum depicted in Fig. 5.15b. Cascading the two resonances appropriately makes our meta-atom exhibit a much enlarged working bandwidth within 2–7 GHz. The physical principle of the tunability can be understood from the equivalent CM of the entire system as shown in Fig. 5.15c, where two magnetic resonances at f_1 and f_2 are modeled

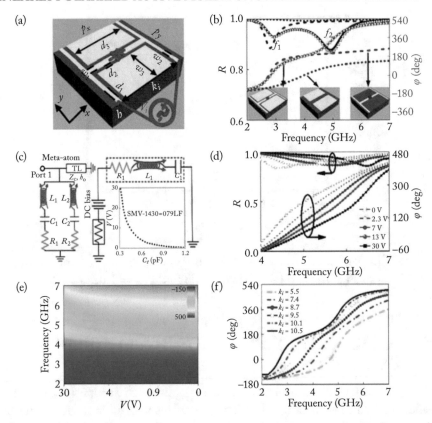

Figure 5.15: (a) Geometry of the basic meta-atom. (b) FDTD (lines) and CM (stars)-simulated spectra of reflectance (black curves, left axis) and reflection-phase (blue curves, right axis) of three periodic metasurfaces with unit-cells being the composite meta-atom (solid lines), meta-atom with patch only (dotted lines), and meta-atom with "H" only (dashed lines), with the capacitance of the varactor diode fixed as $C_t = 0.6$ pF corresponding a fixed biasing voltage 4 V. Here $k_i = 10.5$ mm. (c) The equivalent CM for the tunable meta-atom and the varactor diode SMV1430-079LF with inset depicting its $C_t - V$ curve. (d) FDTD simulated spectra of reflectance and reflection phase of a metasurface consisting of a periodic array of meta-atoms with $k_i = 7.4$ mm, under different biasing voltages. (e) FDTD simulated reflection-phase as a function of biasing voltage V and frequency for the same system studied in (d). (f) FDTD calculated the reflection-phase spectra of periodic metasurfaces with different parameter k_i. The geometrical parameters of the meta-atoms are fixed as $p_x = p_y = 12$, $w_1 = 0.8$, $w_2 = 0.5$, $w_3 = 5.1$, $d_1 = 0.25$, and $d_2 = 0.5$ mm. The circuit parameters in our CM calculations for (b) are $L_1 = 19$ nH, $C_1 = 0.085$ pF, $L_2 = 0.09$ nH, $C_2 = 0.194$ pF, $R_1 = 4.22\ \Omega$, $R_2 = 0.37\ \Omega$, $Z_c = 207.4\ \Omega$, and $h_o = 58.9°$, here Z_c and h_o in CM are the equivalent impedance and electrical length modeling the transmission through the dielectric spacer.

by two series resonant tanks formed by L_1, C_1, and R_1, and L_2, C_2, and R_2, respectively. In our FDTD simulations, the varactor diode is replaced by a series resonant tank with inductance $L_t = 0.7$ nH, resistance $R_t = 1.5$ Ω, and capacitance C_t which sensitively depends on the voltage V imparted on it through a $C_t \sim V$ curve depicted in the inset to Fig. 5.15c. The established CM is justified by the excellent agreement between the results based on FDTD simulations and the CM (see stars in Fig. 5.15b). The physics of the varactor-enabled tunability is thus clear: varying the external voltage V can modify the capacitance C_t of the diode, which in turn, tunes the frequencies (particularly f_1) of the two resonant modes through modifying their effective capacitances C_1 and C_2, and thus the related reflection-phase spectra.

Figure 5.15d compares the FDTD-simulated reflection spectra of a metasurface consisting of a *periodic* array of such tunable meta-atoms, which are simultaneously biased at the same voltage V varying from 0 to 30 V. Indeed, the resonant mode f_1 varies sensitively as a function of V, leading to a dramatic tuning of the reflection phases [207]. However, the reflectance R remains nearly 90% since this is a reflective meta-atom with a continuous metal ground plane and the loss is very small in the microwave regime Fig. 5.15e depicts how the reflection phase φ of the studied tunable meta-atom varies as a function of frequency and the biasing voltage. For every frequency inside the band (4–7 GHz), one can always get a large tuning range of φ (covering more than 175° at ~ 5.7 GHz), which provides large freedom to design our meta-lens.

With the EM properties of the tunable meta-atoms fully understood, we start to design our meta-lenses employing such meta-atoms. Restricted by the geometries of our meta-atoms, we can only realize a one-dimensional *tunable* lens where a column of meta-atoms is controlled by the *same* voltage source. The 2D tunable lens needs sophisticated designs on both the meta-atoms and the control circuits. We assume that our meta-lenses exhibit the following parabolic reflection-phase profiles:

$$\varphi(x, f) - \varphi(0, f) = -\frac{2\pi}{\lambda}\left(\sqrt{x^2 + F^2} - F\right), \tag{5.1}$$

where x denotes the position of the meta-atom and $\varphi(0, f)$ is the reflection phase at the lens center. Here, f and λ are the working frequency and wavelength while F is the designed focal length. We choose 6 different types of tunable meta-atoms (labeled by 1, 2, 3, 4, 5, and 6) to design our meta-lenses, and each vertical row of meta-atoms (all having the same structure) is controlled by an external voltage source (see Fig. 5.16a).

As a starting point, we first design a meta-lens exhibiting the phase profile described by Eq. (5.1) with $F = 60$ mm at $f = 5.5$ GHz, using 6 different types of meta-atoms but under *zero* biasing voltages ($V_i = 0$, $i = 1, \ldots, 6$ and thus $C_t = 1.2$ pF). Without involving too many complications, in our design, we only vary the width of the patch resonator (denoted by k_i for the ith meta-atom) but keeping other geometrical parameters unchanged. Figure 5.15f depicts how the reflection-phase spectrum of our meta-atom varies as its k_i changes. The parameters $\{k_i\}$ were found as 5.5, 7.4, 8.7, 9.5, 10.1, and 10.5 mm, respectively, which can help yield the desired $\varphi(x)$ profile (215.1°, 273.4°, 325.9°, 371.8°, 406.9°, and 429.5°) at $f = 5.5$ GHz. Meanwhile,

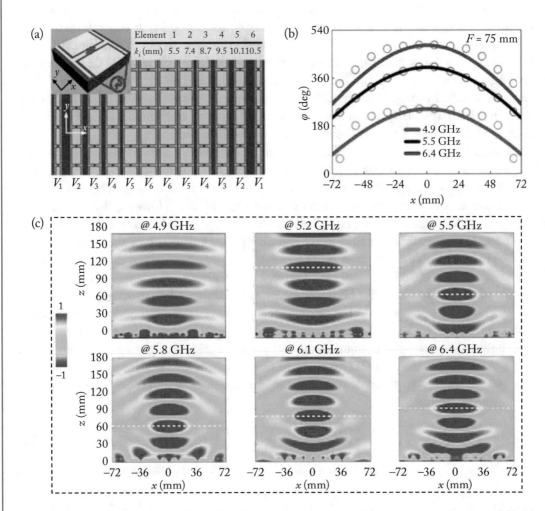

Figure 5.16: (a) Geometry of the designed meta-lens composed by 12×9 meta-atoms. The lens exhibits a $x \rightarrow -x$ mirror symmetry and is formed by six different types of meta-atoms with parameters $k_i = 5.5, 7.4, 8.7, 9.5, 10.1,$ and 10.5 mm. (b) $f(x)$ profiles of the *passive* meta-lens at three different frequencies, obtained by FDTD simulations (symbols) and Eq. (5.1) (lines). (c) FDTD-simulated $Re(E_x)$ field distributions of the *passive* meta-lens at different frequencies when the meta-lens is illuminated by an x-polarized plane wave. The incident field is deducted from the total field. The *passive* lens is taken as the meta-lens under *fixed* biasing voltages corresponding to $F = 75$ mm and $f = 5.5$ GHz.

the reflection amplitudes of all meta-atoms are larger than 0.9 with fluctuations less than 0.1. With the geometrical parameters of these meta-atoms fixed, we then employed FDTD simulations to calculate the $\varphi_i(f) \sim V_i$ relationships for all six meta-atoms (similar to Figs. 5.15d and 5.15e, not shown here). Based on these complete information, we thus retrieved the voltages $\{V_i\}$ imparted on the diodes involved in different meta-atoms employing a root-finding algorithm called traversal query, requiring the meta-lens to exhibit the desired phase profile (Eq. (5.1)) with a pre-determined focal length F at an arbitrary frequency f. The solutions of $\{V_i\}$ depend sensitively on the values of F and f. Since the $\{V_i\}$ solution yielding the desired phase distribution may not be unique, we selected the solution that exhibits the minimum variation on C_t of 6 meta-atoms at each frequency.

To clearly see the dispersion-induced issues in *passive* meta-lenses, we purposely choose a structure with a *fixed* $\{V_i\}$ to *mimic* a passive meta-lens. Without losing generality, here we choose the solution of $\{V_i\}$ obtained for $F = 75$ mm at $f = 5.5$ GHz. Figure 5.16c presents the focusing performances of such a meta-lens at different frequencies. As expected, at $f = 5.5$ GHz the lens does exhibit a perfect focusing performance with a clear focal point at $F = 75$ mm, identified as the position below/above which the EM wavefront is of a concave/convex shape. However, at frequencies other than 5.5 GHz, either the focal point deviates from $F = 75$ mm (5.8, 6.1, and 6.4 GHz) or the focusing performances become bad (4.9, 5.2 GHz), showing that passive meta-lens exhibits clear chromatic aberrations. The underlying physics is very clear. For such a *passive* meta-lens, the intrinsic dispersions of meta-atoms make the phase profiles of the meta-lens deviate quickly from the ideal solution Eq. (5.1), at frequencies other than the target one 5.5 GHz (see Fig. 5.16b).

Such an issue can be solved once we dynamically tune $\{V_i\}$ to rectify the distorted phases and thus make Eq. (5.1) satisfied at *arbitrary* frequencies. We successfully retrieved the requested voltages $\{V_i\}$ at arbitrary frequencies by requiring the resulting $\varphi(x)$ to satisfy Eq. (5.1) with $F = 75$ mm. Figure 5.17a depicts the obtained $V_i \sim f$ curves, which are necessary to correct the otherwise distorted phase distributions (see Fig. 5.17b for $\varphi_i \sim f$ curves under such biasing voltages). As a result, at an arbitrary frequency within the working band, our meta-lens can *always* exhibit the *correct* $\varphi(x)$ satisfying Eq. (5.1) (see inset to Fig. 5.17b for an example at 5.9 GHz). Under these biasing voltages, our meta-lens must be free of aberrations since it can always exhibit perfect phase profiles.

We can further design a functionality-tunable meta-lens utilizing the same approach. Set the working frequency at 5.5 GHz, we retrieved the requested biasing voltages $\{V_i\}$ for the same meta-lens by requiring that its focal length F changes from 45 to 120 mm. Figure 5.17c depicts the obtained $\{V_i\}$ as a function of the designed focal length F. As expected, we have $\{V_i = 0\}$ for $F = 60$ mm, which is the starting point of our design. Figure 5.17d shows how the resultant reflection amplitudes/phases of six meta-atoms vary against F under the voltage combinations $\{V_i\}$ as depicted in Fig. 5.17c. Under the biasing voltages given in Fig. 5.17c, our active meta-lens thus behaves as a functionality-switchable device with tunable focal length.

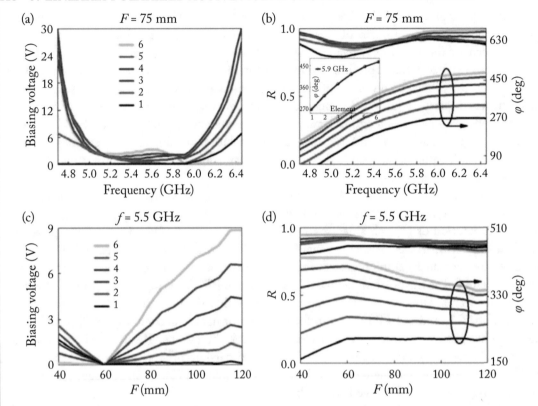

Figure 5.17: (a) Retrieved voltages $\{V_i\}$ and (b) FDTD-calculated reflection coefficients (R and φ) of six meta-atoms as functions of frequency for the designed aberration-free lens with $F = 75$ mm. (c) Retrieved voltages $\{V_i\}$ and (d) FDTD-calculated reflection coefficients (R and φ) of six meta-atoms as a function of the desired focal length F for the functionality-switchable lens working at 5.5 GHz. Inset to (b) shows the phase profile of the meta-lens under the biasing voltages in (a) for $f = 5.9$ GHz.

We fabricated a realistic sample according to the design and experimentally character-ized all the predicted properties (i.e., aberration-free and functionality-switchable imaging). The sample was fabricated based on the printed-circuit-board technique, with an overall size of 144×108 mm^2 (see Fig. 5.18a for its top-view picture). All varactor diodes were attached to the top metallic microstructure using surface-mount technology, which are then biased by appropri-ate constant-voltage sources (GPD4303S). To experimentally test the focusing properties of our meta-lens, we shine the meta-lens (under appropriate biasing voltages as given in Figs. 5.17a and 5.17c) by x-polarized plane waves emitted from a horn antenna placed at 900 mm away from the meta-lens, and then used a 15 mm-long monopole antenna to measure the distribu-

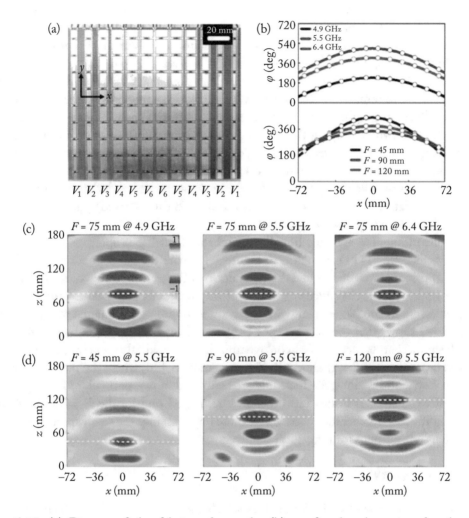

Figure 5.18: (a) Picture of the fabricated sample. (b) φ_x for the aberration-free lens with $F = 75$ mm at three frequencies (top panel), and for the functionality-switchable lens with different focal points ($F = 45, 90,$ and 120 mm) at 5.5 GHz (bottom panel), obtained by FDTD simulations on our tunable lenses (symbols) and Eq. (5.1) (lines). Experimentally measured $Re(E_x)$ distributions for (c) the aberration-free lens with $F = 75$ mm at $4.9, 5.5,$ and 6.4 GHz, respectively and (d) the functionality-switchable lens with $F = 45, 90,$ and 120 mm at 5.5 GHz. The incident field has been eliminated from the measured field. While dashed lines denote the positions of the focal plane.

tions of local E_x field (with both amplitude and phase) on the x-z plane. To see the focusing properties clearly, we purposely deducted the incident field from the measured data so that only the scattered fields are presented.

Figure 5.18 depicts the measured $Re(E_x)$ distributions for our meta-lens under two different groups of biasing voltages as recorded in Figs. 5.17a and 5.17c. As expected, under the biasing voltages as shown in Fig. 5.17a, the meta-lens can always focus incident waves to the focal point at $F = 75$ mm, no matter how we vary the working frequency (see Fig. 5.18c). Such an aberration-free imaging property is in sharp contrast to the passive meta-lens as described in Fig. 5.16c. Meanwhile, keeping the working frequency at 5.5 GHz and adopting the other group of biasing voltages presented in Fig. 5.17c, we found that our meta-lens can exhibit excellent focusing functionalities, but with the focal point switched dynamically from 45–120 mm as we change the biasing voltages appropriately. The intrinsic physics underlying these interesting effects is again that our meta-lens can exhibit the desired (undistorted) phase profiles, under appropriate biasing voltages (see Fig. 5.18b).

The aberration-free and functionality-switchable focusing properties of our meta-lens are finally verified by FDTD simulations. Figure 5.19 plots the corresponding FDTD simulated $Re(E_x)$ distributions of the meta-lens, under two different biasing voltages as shown in Figs. 5.17a and 5.17c. Perfect focusing effects of the meta-lens are observed for all the cases studied, which are in good agreement with their corresponding experimental results. In particular, Fig. 5.19a shows that the performance of our meta-lens can be free of aberrations while Fig. 5.19b shows that its focal lens can be dynamically switched. The good agreement between theory and experiment is clearly shown in Figs. 5.19c and 5.19d, where the measured and simulated field distributions along the focal line are compared in two particular cases studied. The focal-spot size (half-power beamwidth) was measured as 33 mm for the $F = 75$ mm case and increases slightly as F further increases. This is quite physical since the small aperture size makes our meta-lens less efficient for long-focal-length imaging.

5.4   SUMMARY

To summarize, this chapter established a tunable scheme to make multifunctional metadevices in the microwave regime. Adopting tunable meta-atoms involving varactor diodes controlled by external voltages, we can precisely control the phase response of each meta-atom via external knobs, thereby rectifying the inevitable phase distortions at arbitrary frequencies and to encode multiple functions to a single device controlled by external voltages. We experimentally demonstrated two distinct effects: (1) we can realize tunable metasurfaces exhibiting the best functionalities at every frequency within a broadband; and (2) we can dynamically switch the functionalities of metasurfaces via changing the external controls. These results pave the road to realize tunable metadevices achieving dispersion-corrected and/or switchable manipulations of EM waves with high performances, which can stimulate other exciting dynamically switched and dispersion-compensated applications. Extensions to high-frequency domains are even more

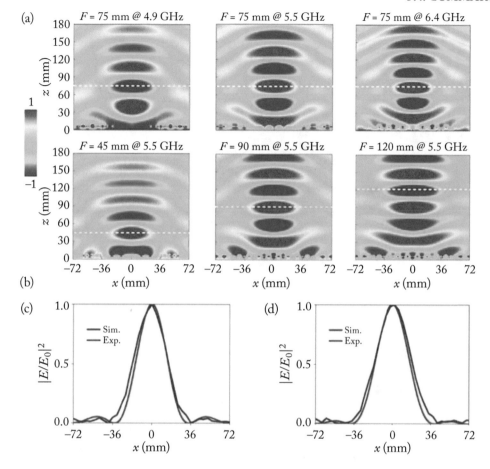

Figure 5.19: FDTD simulated $Re(E_x)$ field distributions for the (a) aberration-free lens (first row) with $F = 75$ mm at 4.9 GHz (left panel), 5.5 GHz (middle panel), and 6.4 GHz (right panel), respectively, and the (b) functionality-switchable lens (bottom row) with F varying from 45 to 120 mm at 5.5 GHz. Comparison between simulated and measured field distributions along the focal line for the tunable meta-lens with (c) $F = 75$ mm and (d) $F = 90$ mm at the frequency 5.5 GHz.

interesting, based on modern technologies such as optical pumping on semiconductors [85] and gate-tunable dielectric materials [205].

CHAPTER 6

Circularly Polarized Active Multifunctional Metasurfaces

In the last chapter, we summarized our efforts on achieving active multifunctional metadevices under the excitations of LP waves. Here, we continue to apply such an active strategy to realize multifunctional metasurfaces/metadevices to control CP waves, yielding fascinating physical effects such as CP helicity conversion and hybridization as well as helicity keeping [81], and dynamically shifted PSHE [208]. The dynamical control over the reflection phase imparts unprecedented capabilities to manipulate the helicity of the incident CP wave. Again, we will start by introducing the design principles of such CP-wave-control metadevices, and then successively present the metadevices that we realized and their functionalities.

6.1 DESIGN PRINCIPLES

Polarization is an important characteristic of electromagnetic (EM) wave [13], and manipulating the polarization states of EM wave is crucial and has attracted intensive attention recently in photonics research. In practical applications, it is highly desirable to have polarization manipulators whose functionalities can be dynamically *tuned* by an external "knob." In the following, we will first establish a general criterion for different helicity states and then propose a strategy in the reflective scheme to realize dynamical control on these helicity states. It is well known that a CP wave will change its helicity after it is reflected back by a conventional mirror. Although in many applications such helicity reversal does not cause substantial problems, in some other cases such as satellite communications, radar detections, etc., it does represent an issue that should be properly solved. Therefore, it is highly desirable to have a tunable device that can dynamically switch its functionality on helicity control of CP waves.

Consider a CP wave with an E-field component $\vec{E}_i = E_0 (\hat{x} \pm i \hat{y}) / \sqrt{2}$ normally incident on a metasurface placed on the xy plane. Suppose the incident wave is propagating along the z-direction, then its polarization is left/right circular polarization (L/RCP) if the sign in front of $i\hat{y}$ is $+/-$, under the time-harmonic convention $e^{i\omega t}$. Assuming that the metasurface exhibits mirror symmetries with respect to the yz and xz planes, the reflected EM wave can be generally expressed as

$$\vec{E}_r = \frac{E_0}{\sqrt{2}} \left(|r_{xx}|e^{i\varphi_{xx}}\hat{x} \pm i |r_{yy}|e^{i\varphi_{yy}}\hat{y} \right), \qquad (6.1)$$

where $r_{xx} = |r_{xx}|e^{i\varphi_{xx}}$ and $r_{yy} = |r_{yy}|e^{i\varphi_{yy}}$ are the reflection coefficients for EM waves polarized along two symmetry axes. To ensure that the reflected wave is still circularly polarized, we require that

$$|r_{xx}| = |r_{yy}|. \tag{6.2}$$

In addition, since the wave-vector of the EM wave is reversed after the reflection, we immediately understand that the handness of the reflected CP wave remains identical to that of the incident one (see Fig. 6.1b) only if the condition

$$\varphi_{xx} - \varphi_{yy} = 180° \tag{6.3}$$

is satisfied. Such a condition is exactly that for high-efficiency photonic spin hall effect (PSHE) derived in Section 6.3. On the other hand, the helicity of reflected CP wave will be reversed (see Fig. 6.1a) if the phases satisfy

$$\varphi{xx} - \varphi{yy} = 0°. \tag{6.4}$$

Note that here the phases are all limited to the branch of $[-180°, 180°]$ without causing any confusion. Finally, when the condition

$$\varphi_{xx} - \varphi_{yy} = \pm90° \tag{6.5}$$

is fulfilled, the reflected wave will become a linearly polarized (LP) beam containing equal amounts of RCP and LCP components; see Fig. 6.1c.

Equations (6.2)–(6.5) show that a metasurface can have diversified helicity-manipulation functionalities if its reflection phases φ_{xx} and φ_{yy} satisfy certain relations. Therefore, we can design a TMS to dynamically control the helicity of EM waves simply by making its φ_{xx} and φ_{yy} satisfy certain desired relations.

Therefore, the basic principle of using metasurface for helicity control is to control its phase response under two orthogonal linear polarizations along its two principal axes, or more relevant to tune the phase differences between them. However, for an active helicity control, it is most often to combine a passive metasurface with active positive intrinsic-negative (PIN) diodes. The underlying physics is that connecting/disconnecting the diode can significantly modulate the EM response (in particular, the reflection phases) of the metasurface via changing its resonance frequency, thereby switch the functionality and operation frequency of the metadevice dramatically. Such a mechanism is fundamentally different from the loss-driven underdamped to overdamped resonator transition [209], since here the reflection amplitude remains nearly unchanged.

6.2   DYNAMICAL CONTROL ON HELICITY OF ELECTROMAGNETIC WAVES

In this section, we design and fabricate a TMS in which a PIN dipole is incorporated into the unit cell, and experimentally demonstrate that the helicity-manipulation functionality of the

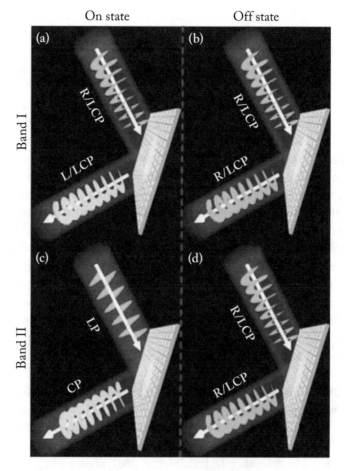

Figure 6.1: Schematics of the proposed multifunctional TMS. The functionalities of the TMS switch between helicity converter and helicity keeper in the band I (a, b) and between LP polarizer and helicity keeper in band II (c, d). Here, the TMS works in either the "On" state (a, c) or the "Off" state (b, d).

device can be dynamically modulated when the diodes are biased at two appropriate voltages. As an example, we employ full-wave simulations to design a frequency-tunable resonant cavity based on our TMS. Our results pave the road to make other tunable and functional devices related to dynamical phase control, such as tunable subwavelength cavities, anomalous reflectors, and even animated holograms.

As an illustration of this general idea, here we present one particular tunable device among many other possibilities. Figure 6.1 schematically depicts what we want to achieve with our device: the metadevice behaves as a helicity converter and a helicity hybridizer for input CP wave within two separate frequency bands in the "On" state, but becomes a helicity keeper (i.e., helicity-conserved reflector) within an ultra-wide band in the "Off" state. In what follows, we describe how to realize such a device.

To realize the functionality-switching effects depicted in Fig. 6.1, we design a TMS whose φ_{xx} and φ_{yy} can be controlled by active elements incorporated in its unit cell. As shown in Fig. 6.2a, our unit cell consists of a metallic electric inductive-capacitive resonator (ELC) coupled with a perfect-electric-conductor (PEC) ground plane through a dielectric spacer, which is the widely available F4B substrate with the permittivity of $\varepsilon_r = 2.65$, the thickness of $h = 6$ mm, and the loss tangent of 0.001. The metallic layers on both sides of the metasurface are copper with a thickness of 36 μm. The ground plane ensures that the device is totally reflective without any transmission. The ELC is basically a topological variant of the split-ring resonator with geometrical parameters shown in Fig. 6.2b. To electrically tune the EM response of the ELC, we connect the two central metallic wires in the ELC resonator by a PIN diode (SMP1345-079LF, Skyworks Solutions Inc. [210]), which can be biased at a given voltage using two high-impedance lines that have perfect electrical connections with the ELC resonator. In the CM shown in Fig. 6.3, the PIN diodes are series-connected lumped elements, which can be modeled as a series of L_s and R_s (L_s and C_j) in the "On" ("Off") state. Here, L_s represents the parasitic inductance, R_s is a total resistance composed of the junction and ohmic resistance, and C_j is junction capacitance. In addition, C_s is the parasitic capacitance in the package and can lead to a little larger C_t (total capacitance) than C_j in the "Off" state. In this particular design, $L_s = 0.7$ nH, $C_j = 0.15$ pF, $C_t = 0.18$ pF, and $R_s = 2\ \Omega$.

To achieve the desired reflection phase and to prevent the microwave signal from entering the bias line, we adopted a Murata LQW04AN10NH00 chip inductor with inductance $L_j = 10$ nH and self-resonance frequency higher than 7 GHz. Applying two appropriate DC voltages on the PIN diode, we can set it working in "On" or "Off" states and thus change the EM responses of the TMS dramatically. Indeed, the TMS shown in Fig. 6.2a can be modeled by appropriate equivalent CM depicted in Fig. 6.2b, at three resonance frequencies. Here, the back metallic layer is represented by ground, whereas the localized magnetic response and the transmission effects through the dielectric substrate (with impedance Z_o and thickness h) were modeled by a transmission line (TL) with equivalent impedance Z_c and electrical length h_o for analysis convenience. According to TL theory, three resonances will occur with central frequencies determined by $f_1^{(y)} = \frac{1}{2\pi\sqrt{L_1C_1}}$, $f_2^{(y)} = \frac{1}{2\pi\sqrt{L_2C_2}}$, and $f_1^{(x)} = \frac{1}{2\pi\sqrt{L_3C_3}}$, respectively. Supported by Fig. 6.6, i.e., $f_2^{(y)}$ originating from the resonance of two symmetric circular arcs and $f_1^{(y)}$ from that of the central wires ("Off" state) or central wire and arcs ("On" state), while $f_1^{(x)}$ from the thin bias line, now the contribution of the loop inductance and capacitance in CM are clear. They are $L_1 \approx L_w + L_s$ and $C_1 = 2C_p + C_f \approx C_f$ in "On" state, and are $L_1 \approx L_w$ and $C_1 \approx C_f * C_t/(C_f + C_t)$ in "Off" state, respectively. Conversely, $L_2 \approx L_a/2, C_2 \approx C_p$ and $L_3 \approx L_j + L_{line}$, $C_3 \approx C_C$ in both states. Here, L_w and L_a are the inductors of the wire and circular arc, and C_p and C_f are the capacitors of the split and the parallel thin lines of two ad-

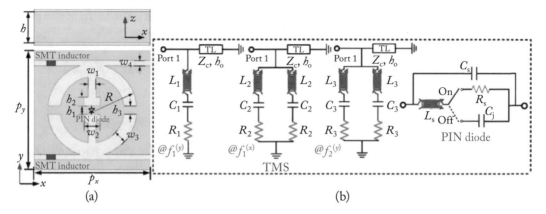

(a)   (b)

Figure 6.2: Topology and equivalent circuit models of the building block of the designed TMS. (a) Side and top views of the building block. (b) Equivalent circuit models of the TMS at resonance frequencies $f_1^{(y)}$, $f_2^{(y)}$ for y-polarization excitation and $f_1^{(x)}$ for x-polarization excitation. Here, $p_x = p_y = 14$ mm, $w_1 = 1$ mm, $w_2 = 2$ mm, $w_3 = 1.5$ mm, $w_4 = 0.6$ mm, $R = 5.4$ mm, $h = 6$ mm, $h_1 = 1$ mm, $h_2 = 1$ mm, $h_3 = 1$ mm, and $L_j = 10$ nH.

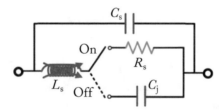

Figure 6.3: Equivalent circuit models of the PIN diode.

jacent elements, and L_j and L_{line} are the inductors of SMT element and thin line, and C_C is the coupling capacitor of the splits between adjacent elements along x-direction. The resonant losses of the above three tanks are represented by R_1, R_2, and R_3, respectively.

Now let us utilize the established criterion to design the dynamic helicity modulator with an optimum bandwidth based on the TL theory. As a starting point, the $ABCD$ matrix of the TMS structure, as well as the equivalent TL, can be expressed as

$$\begin{bmatrix} A_T & B_T \\ C_T & D_T \end{bmatrix} = \begin{bmatrix} 1 & 0 \\ 1/Z_{yi} & 1 \end{bmatrix} \tag{6.6}$$

$$\begin{bmatrix} A_{EqTL} & B_{EqTL} \\ C_{EqTL} & D_{EqTL} \end{bmatrix} = \begin{bmatrix} \cos(kh_o) & iZ_c \sin(kh_o) \\ i \sin(kh_o)/Z_c & \cos(kh_o) \end{bmatrix}, \tag{6.7}$$

where k is the equivalent wave vector of the TEM wave, and Z_{yi} is the impedance of the shunt branch at three resonances and can be calculated as

$$Z_{y1} = i\omega L_1 + \frac{1}{(i\omega C_1)} + R_1, \quad Z_{y2} = i\omega L_2/2 + 1/(2i\omega C_2) + R_2/2$$

and

$$Z_{y3} = i\omega L_3/2 + 1/(2i\omega C_3) + R_3/2,$$

respectively. The total matrix can be readily calculated by cascading the above two $ABCD$ matrices, yielding

$$\begin{bmatrix} A & B \\ C & D \end{bmatrix} = \begin{bmatrix} \cos(kh_o) & iZ_c \sin(kh_o) \\ \cos(kh_o)/Z_{yi} + i\sin(kh_o)/Z_c & iZ_c \sin(kh_o)/Z_{yi} + \cos(kh_o) \end{bmatrix}. \quad (6.8)$$

Then the S parameters with phase information can be immediately achieved from the $ABCD$ matrix by a simple transformation. These analytic expressions are tedious and are not presented here. To obtain a wide operation bandwidth, we require that the reflection phases for two orthogonal polarizations exhibit similar slopes as frequency varies so that their difference can remain nearly a constant within a wide frequency band. Therefore, we enforce the following condition:

$$\frac{\partial \varphi_{xx}(f)}{\partial f} = \frac{\partial \varphi_{yy}(f)}{\partial f} \quad (6.9)$$

at two specific frequencies $f = (f_1^{(y)} + f_1^{(x)})/2$ and $(f_2^{(y)} + f_1^{(x)})/2$ in designing our structure. Here, $f_1^{(x)}$, $f_1^{(y)}$, and $f_2^{(y)}$ are three resonance frequencies which will be explained below. In our CM simulations, the lumped parameters are retrieved in Agilent's Advanced Design System (ADS) from EM simulated reflection magnitude and phase. The CM simulated LP reflection coefficients are then utilized to calculate the CP reflection coefficients using

$$r_{cp} = \begin{pmatrix} r_{RR} & r_{RL} \\ r_{LR} & r_{LL} \end{pmatrix} = \frac{1}{2} \begin{pmatrix} r_{xx} + r_{yy} + i(r_{xy} - r_{yx}) & r_{xx} - r_{yy} - i(r_{xy} + r_{yx}) \\ r_{xx} - r_{yy} + i(r_{xy} + r_{yx}) & r_{xx} + r_{yy} - i(r_{xy} - r_{yx}) \end{pmatrix}.$$

The lumped parameters are retrieved as $L_1 = 30$ nH, $C_1 = 0.107$ pF, $R_1 = 10.9$ Ω, $Z_c = 368.3$ Ω, and $k * h_o = 44.8°$ ranging from 2 to 5 GHz in the "On" state, and as $L_1 = 12.7$ nH, $C_1 = 0.064$ pF, $R_1 = 3.87$ Ω, $Z_c = 371$ Ω and $k * h_o = 38.4°$ ranging from 2 to 5.4 GHz in the "Off" state. As expected, one can see that C_1 and L_1 are significantly altered by the diode while all the other parameters are nearly unaffected between the "On" and "Off" state. These retrieved circuit parameters will be substituted into Eq. (6.9) to check the effectiveness of the design.

Figures 6.4a and 6.4b are the reflection coefficients of the TMS in the "On" and "Off" states, respectively, obtained by FDTD simulations. When the device is in the "On" state (i.e., the PIN diode is forward biased), two resonant modes ($f_1^{(y)} \approx 2.43$ GHz and $f_2^{(y)} =$

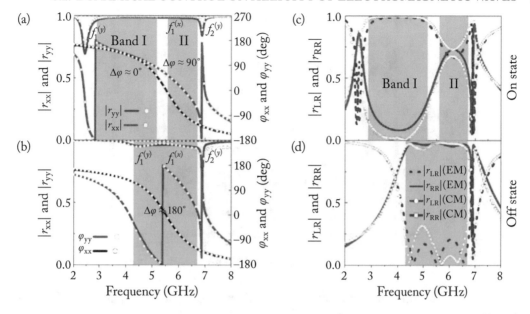

Figure 6.4: Electromagnetic responses of the designed TMS. Calculated LP reflection coefficients (a, b) and CP reflection coefficients (c, d) of the designed TMS in the "On" (a, c) and "Off" (b, d) states based on FDTD (lines) and CM simulation (symbols).

6.88 GHz) can be identified from the calculated $|r_{yy}|$ spectrum, while only a weak resonant dip ($f_1^{(x)} \approx 5.67$ GHz) is found from the $|r_{xx}|$ spectra, see Fig. 6.4a. Meanwhile, $\varphi_{yy} - \varphi_{xx}$ remains nearly 0° within 2.9 ~ 5.13 GHz (Band I) and nearly 90° within 5.58–6.69 GHz (Band II), implying that the device functions as a helicity converter in Band I and as a helicity hybridizer in Band II. Note that the reflection amplitudes $|r_{yy}|$ and $|r_{xx}|$ are both nearly 1 within these two frequency bands. When the device is in the "Off" state, however, its EM responses change drastically; see Fig. 6.4b. Now the low-frequency resonant mode for y-polarization (i.e., $f_1^{(y)}$) is shifted to a new position at 4.6 GHz, while the other two modes are nearly unaffected. As a result, the φ_{yy} spectrum is modified significantly while the φ_{xx} spectrum remains nearly unchanged after the "On-Off" switching. In such a case, $\varphi_{yy} - \varphi_{xx}$ now stays ~ 180° within a broad frequency range (4.33–6.63 GHz, Fig. 6.4b), implying that the device keeps the helicity of incident CP wave after reflection. We note that $|r_{yy}|$ and $|r_{xx}|$ are again nearly 1 within such a band. Such functionality-switching can be more clearly seen from Figs. 6.4c and 6.4d where the FDTD simulated helicity-conserved and helicity-reversed reflectance spectra are shown for the device working in the "On" and "Off" states, respectively. Clearly, the device functions as a helicity converter in Band I and a helicity hybridizer in Band II when it is in the "On"-state, but becomes a CP helicity keeper within a new band (4.33–6.63 GHz) when it is in "Off" state. In

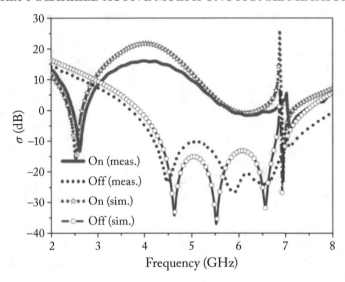

Figure 6.5: Simulated and measured polarization extinct ratio of the TMS plate in the "On" and "Off" state.

both states, the polarization extinction ratio defined as $\sigma = 20\log_{10}(\frac{|r_{LR}|}{|r_{RR}|})$ for the helicity conversion and keeping is more than 10 dB, see Fig. 6.5. In the "On" state, the device "functions as a CP helicity converter with simulated (measured) $|r_{RR}| < 0.31 (|r_{RR}| < 0.29)$, $|r_{LR}| > 0.92$ $(|r_{LR}| > 0.9)$, and > 10.2 dB within 2.9–5.13 GHz (3.11–5.01 GHz), and a CP helicity hybridizer with $\sigma < 3$ dB within 5.58–6.69 GHz (5.51–6.94 GHz). However, the device in "Off" state behaves as a CP helicity retainer with $|r_{LR}| < 0.3 (|r_{LR}| < 0.32)$, $|r_{RR}| > 0.94 (|r_{RR}| > 0.87)$, and < -10 dB within 4.33–6.63 GHz (4.21–7.01 GHz).

To understand the physics underlying such functionality-switching behavior, we studied the current/field distributions of three resonant modes in our structures. Consider the two y-polarization modes first. While in the $f_1^{(y)}$ mode the electric currents are excited to flow in two central vertical lines and two arcs, they are mainly localized along two arcs in the $f_2^{(y)}$ mode (Fig. 6.6). Meanwhile, the x-polarization mode $f_1^{(x)}$ is also predominately associated with the two arcs of the resonant structure, but has nothing to do with the central vertical lines. The nature of these resonant modes helps us understand the principle of the dynamical switching. Since only the $f_1^{(y)}$ mode is connected with the central vertical lines, using a PIN diode to short-circuit the central gap can thus significantly modify such a resonant mode, but has nearly no effects on the other two modes (see Figs. 6.4a and 6.4b). In particular, when the PIN diode is working to connect two central lines (i.e., in the "On" state), the associated resonant structure exhibits a longer metallic length and the induced currents flow in both the central vertical lines and the two arcs. Conversely, while the PIN diode is working to disconnect two central lines (i.e., in the

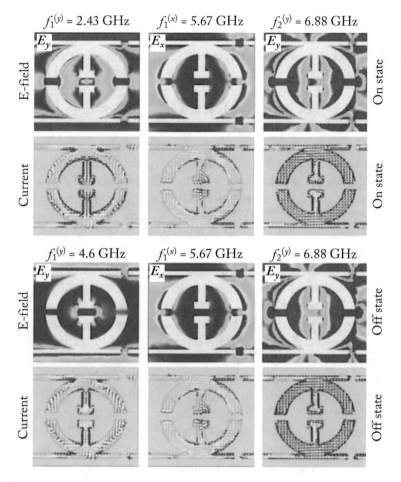

Figure 6.6: The calculated electric field and surface current distributions at three resonance frequencies of the designed TMS.

"Off" state), the associated resonant structure exhibits a shorter metallic length, which explains why the resonance frequency is significantly blue-shifted. Therefore, the working principle of our device is clear: Using a PIN diode to control the resonant frequency of one of the resonant modes, we can dynamically manipulate the phase difference $\varphi_{xx} - \varphi_{yy}$, and in turn, the capability to modulate the helicity of incident CP wave.

More insights can be obtained by checking the equivalent CM of the proposed TMS. As already discussed, the ELC structure possesses two y-polarized resonant modes and one x-polarized mode. Knowledges on these resonant modes help us establish their corresponding equivalent CMs (Fig. 6.2b). The role of the PIN diode is thus clear. The central gap will be connected or disconnected when the PIN diode works in the "On" and "Off" state, respectively, which dynamically controls the resonant frequency $f_1^{(y)}$ via changing the capacitance C and inductance L of the related circuit. In contrast, the other two resonances are nearly unaffected by the PIN diode. To achieve an optimum bandwidth for the proposed device, in real design we fine-tuned our structure to make φ_{xx} and φ_{yy} exhibit similar slopes around the central working frequency in the "Off" state, see established criterion in the previous section. Results calculated with the equivalent CMs are shown as symbols in Fig. 6.4, which are in good agreement with full-wave simulation results.

We fabricated a TMS according to our design and experimentally characterized its EM responses. The TMS is fabricated using the conventional print circuit board technique while the PIN diodes and inductors are attached to the top metallic microstructure using surface-mount technology with leads. The sample consists of 25×25 elements embedded with 625 diodes and 1350 Murata inductors and its total size is 360×360 mm^2 (see Fig. 6.7a). Elements in the same line along x-direction are connected so that they automatically have the same biasing voltage. All the SMT elements are checked by a multimeter of VICTOR VC9807A+ to guarantee the perfect electrical connection and the correct electrodes of the diodes. On the other hand, elements in different horizontal lines are independent and thus their biasing voltages should be carefully adjusted. We note that such a configuration offers us the exciting possibility to independently engineer the phases in different horizontal lines so that the final phase profile can exhibit a linear or parabolic distribution, which can be very useful in other beam-manipulation applications. Here, however, we focus on the situation that all elements are tuned to exhibit the same phase response.

In the measurement setup, see Fig. 6.8, the GPD4303S from GWINSTEK is chosen as DC power and is connected to the two electrodes by the two thin wire. A pair of LP horn antennas (i.e., the source and receiver) are placed at the distance of 1.2 m from the sample to measure the reflection amplitude and phase. Their separation angle is set to be 10° to mimic the normal incidence case in our measurement. By simultaneously changing the orientations of the two horn antennas along x- or y-directions, four LP reflection coefficients are recorded through the Agilent E8362C PNA vector network analyzer, which will be normalized against a reference signal reflected by the metallic plate with the same size of the TMS.

Figures 6.7b and 6.7c depict the experimentally measured reflection coefficients of the fabricated sample, which are in reasonable agreement with FDTD simulation results in Fig. 6.4. The slight differences between simulated and measured spectra are primarily due to the additional resistances from the soldering pads, and the misalignment of the source/receiver antennas. As expected, three resonant dips appear at $f_1^{(y)} = 2.6$ GHz, $f_2^{(y)} = 6.99$ GHz, and $f_1^{(x)} = 5.58$ GHz when the device is in the "On" state and their positions are shifted to $f_1^{(y)} = 4.65$ GHz, $f_2^{(y)} = 7.03$ GHz, and $f_1^{(x)} = 5.62$ GHz when the device is in the "Off" state. In addition, the shallow dip and slowly dispersed phase response at $f_1^{(x)}$ indicate that such a mode exhibits a low-quality factor, as expected. However, compared with the FDTD spectra (Fig. 6.4), now the measured reflection phase only covers a small region for the mode $f_2^{(y)}$. This can be attributed

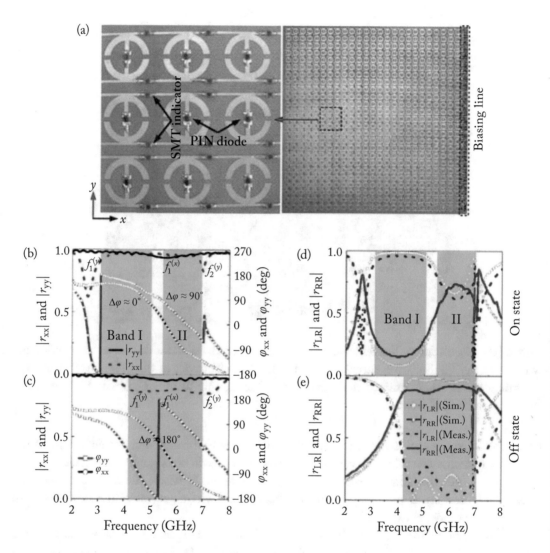

Figure 6.7: Picture and the measured LP and CP reflection coefficients of the fabricated TMS. (a) The building block of the TMS is connected by the SMT inductors along x-direction with the biasing line to control the voltage of the PIN diode. (b, c) In LP basis, the reflection phase difference between φ_{yy} and φ_{xx} is kept at about $0°$ in Band I (3.11–5.01 GHz) and about $90°$ in Band II (5.51–6.94 GHz) in the "On" state and is kept at about $180°$ in an ultra-wideband (4.21–7.01 GHz) in "Off" state. (d, e) In the CP basis, the TMS functions as a CP helicity converter in Band I and CP helicity hybridizer (i.e., CP polarizer) in Band II in the "On" state, and the CP helicity keeper in the ultra-wideband (4.21–7.01 GHz).

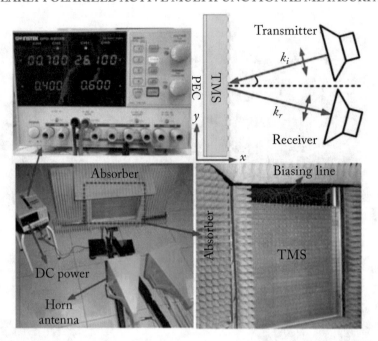

Figure 6.8: Illustration of the reflection measurement setup. All cathode and anode conducting wires from the TMS plate eventually pooled in the two electrodes of the DC power. A mass of absorbing materials are utilized aside from the TMS sample to fully eliminate the reflections from surrounding objects.

to the enhanced absorption in the real sample which destroys the high-quality-factor resonance, see detailed discussions below. Nevertheless, such deviations do not significantly affect the functionality of our device. Indeed, as shown in Figs. 6.7d and 6.7e where the reflection spectra in terms of CP bases are re-plotted, the functionality of our device changes significantly as we turn "On/Off" the diode, as expected.

To understand the deviations between the simulated and measured reflection phase around $f_2^{(y)}$ (see Figs. 6.4 and 6.7), we have performed more FDTD simulations to investigate the influence of the resistance R_j of the SMT elements on the reflection phase. Such an effect can be well explained by the coupled-mode theory (CMT) [211–213], based on which the complex reflection coefficient r of the metasurface can be derived as:

$$r = -1 + \frac{2/\tau_r}{-i(\omega - \omega_0) + 1/\tau_a + 1/\tau_r}. \tag{6.10}$$

Here, τ_a and τ_r are the lift times of the resonance due to absorption inside the structure and radiation to the far field, respectively. According to CMT, the absorptive quality factor $Q_a = \frac{\omega_0 \tau_a}{2}$ and the radiation quality factor $Q_r = \frac{\omega_0 \tau_r}{2}$, retrievable from the simulated or measured

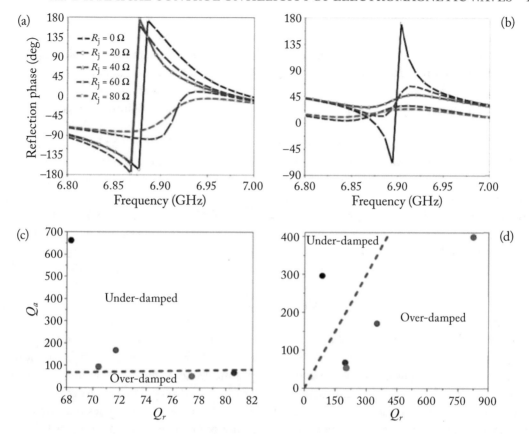

Figure 6.9: The reflection phases (a, b) and the calculated $Q_a \sim Q_r$ phase diagrams (c, d) of the TMS plate in the "On" (a, c) and "Off" (b, d) states. The dashed line in (c) and (d) correspond to that of $Q_a = Q_r$.

reflection spectra, will collectively determine the metasurface to be either an EM absorber or an EM phase modulator.

Here, we study the influence of R_j on the reflection phase of TMS based on FDTD simulations. As shown in Figs. 6.9a and 6.9b, the reflection phase undergoes a continuous variation from $-180°$ to $180°$ while R_j increases from 0 to 40 Ω in the "On" state. However, if R_j increases to 60 Ω or 80 Ω, the reflection phase can only cover a range of less than 90°. Such behaviors are consistent with the Q diagrams (Fig. 6.9c), showing that Q_r and Q_a are located in the under-damped region (above the red dashed curve) when $R_j = 0$, 20, and 40 Ω, while in the over-damped region (under red dashed curve) when $R_j = 60$ and 80 Ω. Similar arguments are also valid in the "Off" state. As shown in Figs. 6.9b and 6.9d, the TMS is located in the under-damped region showing a full range phase (i.e., 360°) modulation as $R_j = 0$ Ω, while in the over-damped region showing a limited phase modulation as $R_j = 20, 40, 60,$ and 80 Ω. Therefore, the inconsistency between the simulated and measured reflection phase around f_2 may attribute to the increased R_j in the real fabricated sample. To demonstrate this argument, we have retrieved the absorptive quality factor ($Q_a = 31.9$ Ω) and the radiation quality factor ($Q_r = 339.2$ Ω) according to the measured reflection spectra, indicating that the TMS is located in the over-damped region (i.e., $Q_a < Q_r$) and thus supplies a limited reflection phase region.

Our TMS can find many other interesting applications, which will be briefly introduced in this section. For example, in previous discussions, we only used the diode to control one of the resonant modes. Obviously, more fascinating functionalities can be realized if we choose to control two or three resonant modes simultaneously. In addition, here we only utilized the $\Delta \varphi$ degree of freedom to achieve the polarization modulation, but in fact, individually tuning φ_{yy} or φ_{xx} can also find many interesting applications. As schematically shown in Fig. 6.10a, we can combine a TMS with a passive reflector to form a double-plate resonant cavity. According to [214], now the resonance frequencies of such a metamaterial-based cavity are determined by

$$\varphi_1 + \varphi_2 - \frac{4\pi f d}{c} = 2n\pi, \tag{6.11}$$

where φ_1 and φ_2 are, respectively, the reflection phases of EM waves at the TMS and the passive reflector, d is the distance between two plates and n is an arbitrary integer. Different from a conventional cavity, now φ_1 of our TMS can be dynamically controlled by switching on/off the diode, and thus the resonance frequencies of our cavities can also be dynamically controlled. In addition, the fact that φ_1 of our TMS covers the whole 2π phase range implies that the size of our cavity can break the half-wavelength restriction imposed on conventional cavities [214].

We used FDTD simulations to demonstrate the feasibility of the proposed idea. In our design, the passive reflector consists of complementary ELCs (CELCs) printed on an F4B dielectric substrate (thickness $h = 1.5$ mm, $\varepsilon_r = 2.65$ with loss tangent of 0.001), and the cavity thickness is set as $d = 4$ mm. We first studied the reflection/transmission spectra of the passive reflector. Figure 6.10b shows that the passive reflector is highly reflective for y-polarized EM waves with the reflection phase staying around 180° in the frequency range of interest (2– 7 GHz), indicating that it is just a conventional electric reflector. To probe the resonant modes of our cavity, we put a y-oriented 6-mm-long dipole antenna into the cavity and use FDTD simulations to study its radiation properties. Figures 6.10c and 6.10d show that the S_{11} spectra of the antenna are dramatically changed after it is placed into the cavity working in "On" and "Off" states, respectively. In particular, the new pronounced dips (around 2.62 GHz in "On" state, and around 4.86 GHz in "Off" state) in S_{11} spectra are the signature of the resonance modes, since such a short antenna does not radiate efficiently at these frequencies (see the blue curve for S_{11} spectra of the bare dipole antenna). We note that the thickness of our cavity is only about $\lambda_0/30$ and $\lambda_0/15$ at these two resonance frequencies, indicating that our cavity is deeply subwavelength. As a comparison, the lowest resonance mode is at 37.5 GHz for a conventional

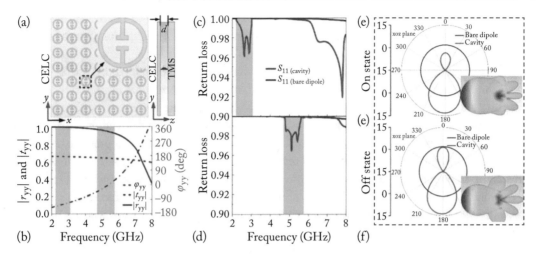

Figure 6.10: The schematics, reflection coefficients, and resonance modes of the frequency-tunable subwavelength cavity. (a) The schematics of the cavity consisting of CELC and TMS. (b) Scattering coefficients of the CELC, (c, d) return loss spectrum, and (e, f) radiation pattern of the subwavelength cavity in the "On" and "Off" state. The lateral size of both CELC and TMS plate is 182 mm × 182 mm in the "On" state while 98 mm × 98 mm in the "Off" state. The geometrical parameters of the CELC structure are $p_x = p_y = 14$ mm, $w_1 = w_3 = h_2 = 1$ mm, $w_2 = 3$ mm, $R = 5$ mm, and $h = h_1 = 1.5$ mm.

double-plate cavity with the same thickness. Most importantly, the working frequency band of our subwavelength cavity is dynamically switched by controlling the working state of the PIN diodes incorporated in the TMS. The underlying physics is that turning on/off the PIN diode can dynamically control φ_{yy} of the TMS, and in turn, switch the working frequencies of the resonant modes through the phase-matching condition. Figures 6.10e and 6.10f depict the radiation patterns of the cavity working in "On" and "Off" states, respectively. As expected, now our subwavelength cavity exhibit unidirectional radiations with narrowed half-power beamwidths. The directivity is increased by more than 10.3 dB and cannot be inspected at other dips which do not correspond to highly directive emissions.

6.3 TUNABLE PANCHARATNAM–BERRY METASURFACES

PB metasurfaces, with orientational rotations and CP wave stimulations, have intrigued a great deal of interest in recent years. Along with the success of ushering PB metasurfaces in anomalous reflection/refraction, the vast potential applications to novel devices are also evidenced due to their broad operation band and versatile features, such as vortex plate [215], orbital angular momentum [216], ultrathin flat lens [48, 161], PSHE [55], contoured beam synthesis [217, 218], beam steering antenna [219, 220], high-gain lens array [221–226], holograms [43], and so on.

Unfortunately, above PB metasurfaces or devices are confined to fixed electrical performance/functionality once the design is accomplished, indicating somewhat limited freedom, low reusability and integrity, and a waste of resources. Moreover, their functionality cannot be dynamically modulated with versatile performances despite a broadband dispersionless phase gradient. Such two issues would hamper the applications of PB metasurfaces in practice. Although tunable approaches have been investigated recently to control the EM response of metasurfaces in different frequency domains, they are mostly confined to homogenous metasurfaces without geometry rotation [208]. To date, tunable *inhomogeneous* PB metasurface with frequency and/or functionality reconfigurability is rarely seen due to the lack of efficient feeding technique for rotating structures. Moreover, tunable metasurface typically features limited phase and frequency tuning range, sharp resonant dips and a small phase shift less than 180°. This is because tunable diodes commonly introduce LC elements and undesired resistive loss, which enhance considerably the quality (Q) factor and absorptions. As a result, sharp dips and swift phase shifts are observed for reflection magnitude and phase response, yielding identical phases (asymptotic behavior) near below and near above the resonance.

Here, we report a strategy for PB metasurface with agile working frequency by involving each meta-atom with tunable PIN diodes. To bridge the gap between flexible functionality and dynamical phase control, we conceived an efficient feeding strategy by proposing a new topology of a slip ring brush. Such a strategy enables the active feeding of every rotated meta-atoms by involving each of them with a PIN diode. Of particular relevance is the mechanism studied for suppression of normal reflection modes from the viewpoint of cascaded Jones matrix. Our proposal, not confined to PHSE, set a solid platform for PB phase control and can be populated to any helicity-trigged dual-functional and/or multifunctional devices with high integrity, stability and low cost.

To accomplish our target, the basic meta-atom should exhibit dual-mode operations in the "Off" state while a broadband single-mode operation in the "On" state. One easy and straightforward way is to design dual-mode meta-atom in the "Off" state and then close up one of the dual modes by biasing PIN diodes to "On" state. Figure 6.11 depicts the topology of the proposed meta-atom as a basic building block of the tunable PB metasurface. As can be seen, the meta-atom is composed of three metallic layers sandwiched by two dielectric spacers: the top composite metallic pattern, the middle ground plane, and the bottom feeding network. This configuration enables a completely reflective system without transmissions. In practice, the two dielectric spacers can be bonded together by adhesives and reinforced by hot press. To break the asymptotic behavior and thus facilitate an active design, the meta-atom is required with a sufficiently large phase and frequency tuning range. In this regard, the top metallic pattern is purposely designed as a composite of "I" structure and a pair of symmetric patches. When

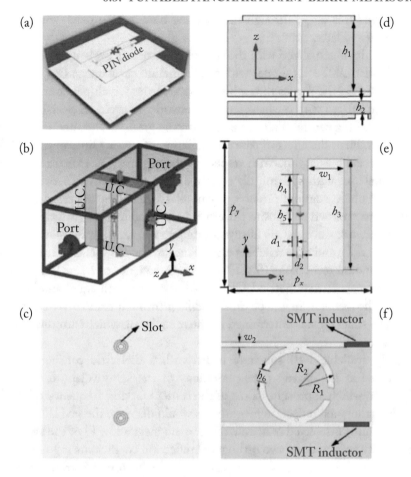

Figure 6.11: Topology of the proposed tunable PB meta-atom. (a) Perspective view. (b) Simulation setup. (c) Slot in the ground. (d) Side view. (e) Top view. (f) Bottom view.

illuminated by normally incident y-polarized transverse EM wave along the z-direction, the electric and magnetic fields drive the "I" structure and patch, and each couple to the ground and generates an independent magnetic resonance around $f_1^{(y)}$ and $f_2^{(y)}$, respectively. Our dual-mode strategy is analogous to the V-antenna by combining its two eigenmodes but is with an additional degree of freedom for individual control of $f_1^{(y)}$ and $f_2^{(y)}$. Such a multi-resonant element feature relaxes the high Q of a resonant circuit and thus efficiently resolves the issues of sharp resonance and limited phase/frequency tuning range in previous tunable work. Note that two different "I" structures kept side by side also enable similar dual magnetic resonances, however, the simultaneous control of them requires two diodes and two different biasing lines.

Our strategy enables this target through the coupling between the patch and the "I" structure which will be discussed later.

Here, tunability is actualized with the central "I" structure broken and then connected by M/A-COM MA4PBL027 PIN diodes with small junction capacitance C_j [227]. The external voltages are imposed on the diodes through two metallic vias which penetrated the two dielectric spacers and terminated with the back feeding network. To avoid an electrical DC short, two circular slots each with a central plate are etched on the ground. The functions of the metallic ground are twofold. First, the effect of parasitic radiation of feeding lines to the PSHE is inhibited. Second, the fixed biasing line with respect to the top metallic structure does not destroy the two orthogonal axes with dynamically rotated angles. Otherwise, the orthogonal axes disappear and the PB phase does not preserve anymore. To guarantee the electrical connection of DC source to the top rotated composite pattern, an electrical slip ring brush is proposed as the bottom feeding network since the via always located on the slip ring no matter how it rotated. To provide an RF choke and thus reliable performance of the PB metasurface, Murata chip inductors in surface mount technology (SMT) with the inductance of $L_j = 10$ nH are introduced in biasing lines. Moreover, symmetric LCP and RCP beams with equal amplitude will be restored, for which the mechanism will be explained later through field/current distributions. The required extremely small C_j was determined by the metallic via which introduces undesired loss and large inductance.

The widely available F4B substrate board with a dielectric constant of $\varepsilon_r = 2.65 + 0.00265i$ is utilized as the spacers. The periodicity of the subwavelength square element is $p_x = p_y = 12$ mm which approaches $\lambda_0/3$ at the center working frequency of 10 GHz. Since the nonuniform orthogonal reflection magnitudes would decrease the PSHE efficiency, the top substrate ($h_1 = 3$ mm) is selected relatively thick to engineer a low Q of the system. In this regard, the near-unity amplitude of two orthogonal reflection coefficients is guaranteed, and thus large frequency and phase tuning range are expected. The bottom substrate for supporting the biasing circuit poses little effect to the EM response due to the shielding of the ground and is designed as $h_2 = 0.5$ mm for the sake of a thin structure.

In the equivalent CM under y polarization, see Figs. 6.12a–d, the single magnetic resonance at $f^{(y)}$ in the "Off" state is modeled by L_2, C'_2, and R'_2, whereas in the "On" state the two magnetic resonances at $f_1^{(y)}$ and $f_2^{(y)}$ are associated with I structure and patch and are modeled by two series resonant tanks formed of L_1, C_1, R_1 (resonant loss), and L_2, C_2, R_2 (resonant loss), respectively. The middle metallic layer was represented by the ground while the reflection through the dielectric was modeled by a TL with effective impedance $Z_c = \frac{Z_0}{\sqrt{\varepsilon_r}}$ and electrical length $h_o = 2h_1 - \frac{\lambda_0}{(2\sqrt{\varepsilon_r})}$, where Z_0 and λ_0 are the wave impedance and wavelength in vacuum. For x polarization (Fig. 6.12e), the magnetic resonance originated from the coupling between patch and ground is modeled by a resonant tank formed of L_3, C_3, and R_3 in both states. Such a magnetic response for both orthogonal polarizations is determined by the line inductor L_{p1} (L_{p2}) and parallel-plate capacitor C_{p1} (C_{p2}) of the patch. By

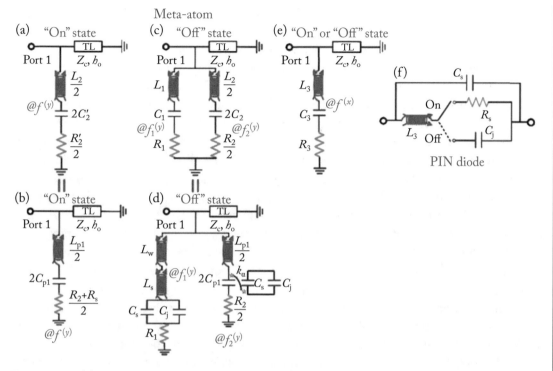

Figure 6.12: The equivalent circuit model of the (a–e) proposed meta-atom for (a–d) y polarization and (e) for x polarization at $f^{(x)}$ and (f) microwave model of PIN diodes. The (a, c) general and (b, d) detailed model and of the meta-atom at (a, b) $f^{(y)}$ in the "On" state and at (c, d) $f_1^{(y)}$, $f_2^{(y)}$ in the "Off" state.

controlling the voltages imposed on the diodes, the working state of the PIN diode can be switched between "On" and "Off" state which corresponds to a series of lead inductance L_s and resistance R_s, and a series of L_s and C_j, respectively, in Fig. 6.12f [227]. The association of various lumped elements with circuit parameters in "On"/"Off" state under x/y polarizations is now clear: $L_1 \approx L_w + L_s$, $L_2 \approx L_{p1}$, $L_3 \approx L_{p2}$, $R'_2 = R_2 + R_S$, $C_1 = C_s + C_j$, $C_2 \approx C_{p1} + k_\alpha * (C_s + C_j)$, $C'_2 = C_{p2}$, and $C_3 \approx C_{p2}$, where L_w is bar inductor, C_s is package capacitance, and k_α measures the coupling between the patch and I. Therefore, the resonances of the meta-atom under y polarization and thus the phase response at a fixed frequency can be arbitrarily tuned in terms of varied LC values. Note that a progressive control can be implemented by utilizing varactor diodes with a significant tuning range of C_j. However, such a progressive control is hard to realize by the high-impedance meta-atom with metallic vias. This is because such element topology requires extremely small C_j which typically exhibits small variation and cannot be afforded by commercially available tuning varactors.

For characterization, the unique EM behavior of the meta-atom in the "Off" state ($L_s = 0.11$ nH, $C_j = 0.03$ pF) is evaluated in CST Microwave Studio where the unit cell boundary is assigned to four walls along x and y directions to mimic an infinite array, see Fig. 6.11b. For a comprehensive study, five cases are considered. As shown in Figs. 6.13a and 6.13b, there is only a single resonant mode identified from the reflection dip and sharply changed phase in either bare I structure or bare patch case. However, with both patch and I structure involved, two magnetic resonances are observed at the lower ($f_1^{(y)} = 5.8$ GHz) and upper ($f_2^{(y)} = 8.7$ GHz) frequencies when $h_3 = 9$ mm, where the phase changes rapidly across $0°$ twice. The reflection response calculated from EM and CM simulations coincides well, revealing the rationality of the model. The $f_2^{(y)}$ shifts upwards as h_3 decreases and therefore we do not observe the high-frequency resonance within the available band when h_3 is 3 and 6 mm. The dual-mode resonances are in much proximity in reflection spectra when h_3 is sufficiently large. As a result, the dual-mode operation breaks the asymptotic phase at edge frequencies, yielding a large phase tuning range and a near-unity reflection amplitude. Moreover, the coupling between the patch and I induced slightly lower $f_1^{(y)}$ and higher $f_2^{(y)}$ in the composite meta-atom. The origin of $f_1^{(y)}$ from I structure while $f_2^{(y)}$ from the patch is further illustrated from the surface current distributions shown in the insets to Fig. 6.13b, where strong current intensity is observed on I structure and patch at f_1 and f_2, respectively. Although strong current intensity is also observed on I structure at f_2, it is in reversed phase with that on patch due to the parasitic coupling from the patch.

Figures 6.13c and 6.13d depict the reflection coefficients of the tunable meta-atom for different C_j when other geometrical parameters are kept constant while Figs. 6.13e and 6.13f depict those for different C_1 obtained from CM simulations. As is expected, the trends predicted from CM are in good consistency with those from EM results. Both $f_1^{(y)}$ and $f_2^{(y)}$ undergo a red-shift when C_j increases from 0.03 to 0.12 pF or C_1 from 0.06 to 0.12 pF. The capacitive coupling between the patch and I structure enables such a simultaneous control. Moreover, the reflection dip at $f_1^{(y)}$ reduces remarkably with C_j and C_1. This is because the increase of capacitance enhances the energy stored in the circuit and thus reduces the wave intensity scattered in free space. The non-uniform amplitude would induce undesired scattering modes of PB metasurface and thus degrade the PSHE efficiency. This EM feature makes the progressive control through varactors inappropriate but the discrete control through PIN diodes more suitable. To validate this proposal, we also calculate the reflection spectrum in the "On" state ($L_s = 0.11$ nH, $R_s = 2.8\ \Omega$); see Figs. 6.13g and 6.13h. As can be seen, the resonant response of the "I" structure dies off and a new resonance $f^{(y)}$ occurs between $f_1^{(y)}$ and $f_2^{(y)}$. The origin of $f^{(y)}$ from the patch can be validated from the progressive redshift with h_3, and the almost identical $f^{(y)}$ in patch only case. The changed and shifted resonances in both states afford considerable phase agile which is the key to implement phase-dependent devices with dynamically switched functionality. This can be implemented by engineering the biasing circuit forward or zero biased through imparted DC voltages so that the PIN diodes work in the "On" or "Off" state.

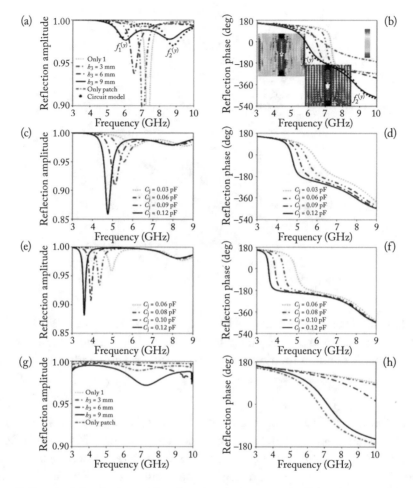

Figure 6.13: Effects of the (a, b, g, h) patch height h_3 and (c–f) junction capacitance C_j on reflection response of the tunable meta-atom under y polarization. EM-calculated reflection response in the (a–d) "Off" and (g, h) "On" state. (e, f) CM calculated the reflection response in the "Off" state. (a, c, e, g) Reflection amplitude. (b, d, f, h) Reflection phase. For general purposes, we plot here the results for meta-atom without lumped inductors in the bias line since they are the same as those for meta-atom with lumped inductors when self-resonance of the lumped inductors does not occur. Note that $h_3 = 9$ mm was used for the "only patch" case. The geometrical parameters are designed as (unit: mm) $p_x = p_y = 12$, $d_1 = 0.4$, $d_2 = 0.5$, $d_3 = 0.3$, $w_1 = 3$, $w_2 = 0.4$, $R_1 = 3$, $R_2 = 2.5$, $h_1 = 3$, $h_2 = 0.5$, $h_3 = 9$, $h_4 = 2.5$, $h_5 = 1$, and $h_6 = 1$. In full-wave simulations, $L_s = 0.115$ nH, $C_j = 0.03$ pF, and $R_s = 2.8$ Ω. The circuit parameters are retrieved as $L_1 = 13$ nH, $C_1 = 0.04$ pF, $L_2 = 1$ nH, $C_2 = 0.081$ pF, $R_1 = 0.9$ Ω, $R_2 = 0.8$ Ω, $Z_c = 109.3$ Ω, and $h_o = 60.5°$.

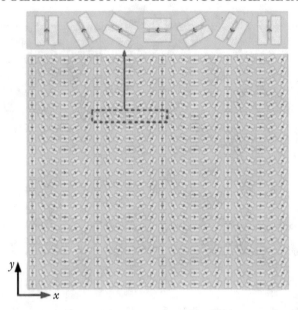

Figure 6.14: Layout of the designed tunable PB metasurface with the zoom-in view of a supercell shown in the dashed. The geometrical parameters of the meta-atoms are the same as those shown in Fig. 6.13 except for different rotation angles and introduced chip inductors.

With all fundamentals and EM behavior of the meta-atom known, we now describe our strategy to design a functionality switchable PB metasurface utilizing these basic building blocks. Figure 6.14 portrays the layout of the designed PB metasurface. As is shown, the metasurface is composed of several supercells periodically arranged along x- and y-direction, respectively. The supercell consists of six ($N = 6$) sequentially rotated meta-atoms with identical geometrical parameters in a rotation angle step of $\Delta\varphi = 30°$ along the x-direction, such that they provide an incremental phase shift of $\pi/3$ and exhibit desired phase gradient following the relationship $\xi = \frac{\pm 2\pi}{(Np_x)}$. To guarantee an efficient biasing for meta-atoms rotated by 90°, the splits introduced in the ring are purposely rotated by 15° in the initial element; see Fig. 6.11f. According to generalized Snell's law, the scattered patterns from the normally incident CP wave will be directed toward off broadside by $\theta_r = \arcsin(\lambda\xi/2\pi)$ with respect to the normal. Therefore, two split EM beams are expected traveling along with inverse directions if the metasurface is illuminated by a normally incident LP wave which can be represented by the sum of an RHCP wave and an LHCP wave. Note that there are also other scattered modes (normal reflections) which will inevitably decrease the PSHE efficiency.

Here, we will briefly analyze the condition for high-efficiency PSHE below based on the scattering matrix method [224]. For a reflection meta-atom under the Cartesian coordinate system, the Jones matrix in reflection scheme after a rotation angle φ can be written as

$$R_{\varphi}^{XY} = \begin{bmatrix} \cos\varphi & -\sin\varphi \\ \sin\varphi & \cos\varphi \end{bmatrix}^{-1} \begin{bmatrix} r_{xx} & r_{xy} \\ r_{yx} & r_{yy} \end{bmatrix} \begin{bmatrix} \cos\varphi & -\sin\varphi \\ \sin\varphi & \cos\varphi \end{bmatrix}, \tag{6.12}$$

where the superscripts x and y denote the polarization of incidence. Then we can immediately obtain the reflection matrix under a CP basis:

$$R_{\varphi}^{LR} = \frac{1}{2} \begin{bmatrix} 1 & -j \\ 1 & j \end{bmatrix} R_{\varphi}^{XY} \begin{bmatrix} 1 & -j \\ 1 & j \end{bmatrix}^{-1}. \tag{6.13}$$

By substituting Eq. (6.12) into (6.13) we derive the following scattering matrix under CP incidence after some algebraic manipulations:

$$r_{ll} = \frac{1}{2}[(r_{xx} - r_{yy}) - j(r_{xy} + r_{yx})]e^{-j2\varphi} \tag{6.14a}$$

$$r_{rr} = \frac{1}{2}[(r_{xx} - r_{yy}) + j(r_{xy} + r_{yx})]e^{+j2\varphi} \tag{6.14b}$$

$$r_{lr} = \frac{1}{2}[(r_{xx} + r_{yy}) + j(r_{xy} - r_{yx})] \tag{6.14c}$$

$$r_{lr} = \frac{1}{2}[(r_{xx} + r_{yy}) - j(r_{xy} - r_{yx})]. \tag{6.14d}$$

Equations (6.14) show that for rotated meta-atoms, only two components r_{ll} and r_{rr} carry the PB phase information, which is twice the rotation angle [156], while the other two components r_{lr} and r_{rl} do not supply any PB phase and contribute to normal reflections. To ease the design, we design our system without chirality and thus without cross-polarization conversion under LP illumination ($|r_{xy}| \approx |r_{yx}| \approx 0$). To eliminate the undesired scattering modes, we only require $r_{xx} + r_{yy} = 0$. In a reflection scheme, we can easily obtain two orthogonal near-unity reflection amplitudes ($|r_{xx}| \approx |r_{yy}| \approx 1$). Therefore, we only need $\varphi_{xx} - \varphi_{yy} \approx 180°$ to guarantee a high PSHE efficiency of our system. Such phase relation has been successfully utilized for polarization control and highly directive subwavelength cavity [208]. By following the developed dispersion engineering method based on Eq. (6.9), the 180° phase difference between two orthogonal reflections of our meta-atom can be achieved in broadband in the "On" state. Therein, we require that the reflection phases for two orthogonal polarizations exhibit similar slopes at several test frequencies. Generally speaking, these test frequencies can be arbitrarily selected and are chosen as $f = (f^{(y)} + f^{(x)})/2$ and $f = f^{(x)}$ in this work for easy design without loss of generality. In our design w_1 and h_3 are chosen to optimize the dispersion curve through iterative full-wave simulations.

Figures 6.15a and 6.15b show the two orthogonal reflection coefficients of our finally designed meta-atom in both "On" and "Off" states. As is expected, two well-separated resonant dips are clearly observed around $f_1^{(y)} \approx 5.62$ and $f_2^{(y)} = 8.45$ GHz where the phases

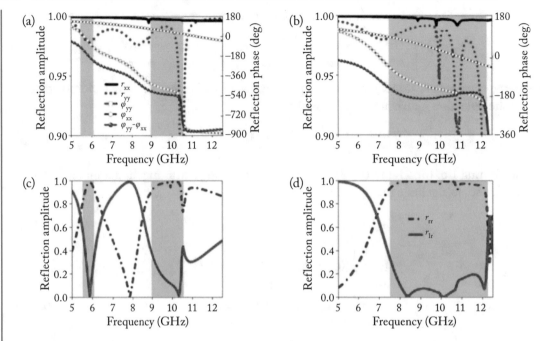

Figure 6.15: Simulated (a, b) LP and (c, d) CP reflection coefficients of the designed PB meta-surface in the (a, c) "Off" and (b, d) "On" states.

reach zero degree in the "Off" state from $|r_{yy}|$ spectrum. The $|r_{yy}|$ is more than 0.97 at most observed frequencies of 5–12.5 GHz except for that around 10.5 GHz where a parasitic resonance of I structure through the feeding network occurred. However, there is only a resonant dip around $f^{(y)} = 7.01$ GHz in the "On" state where the phase reaches 0°. Moreover, the band for $|r_{yy}| > 0.9$ ranges from 5 to 12 GHz. The self-resonance of lumped inductors and parasitic resonances of I structure and feeding network give rise to three shallow dips after 10 GHz. In both "On" and "Off" states, only a weak resonant dip occurs around $f^{(x)} = 11.2$ GHz from $|r_{xx}|$ spectra. The out-of-phase difference $(\varphi_{yy} - \varphi_{xx})$ with a tolerance of $-180° \pm 40°$ is obtained within 5.67–6.08 GHz (Band I) and 9–10.55 GHz (Band II) in the "Off" state, whereas it is observed within 7.5–12.2 GHz (Band II) in the "On" state. Therefore, a dual-band and a broadband high-efficiency PSHE is expected in the former and latter cases, respectively.

Figures 6.15c and 6.15d further depict the CP reflection coefficients of the PB metasurface under RCP wave illumination. As is expected, slight ripples observed at high frequencies of the spectrum are attributable to the parasitic resonances for both orthogonal polarizations. The non-uniform reflection amplitude and fluctuated reflection phases give rise to the shallow dips in r_{rr} the spectrum at 10.5 GHz in the "Off" state while at 9.9 and 10.6 GHz in the "On" state. Nevertheless, the variation is at an acceptable level such that the broadband was not divided into two in the "On" state. The bandwidth characterized by $|r_{lr}| < 0.3$ (-10 dB) and $|r_{rr}| > 0.9$ is 7.6–12.1 GHz in the "On" state while it is 5.7–6.05 GHz, and 9.15–10.5 GHz in the "Off" state.

For demonstration, we have performed extensive characterizations on far-field scattering patterns of the PB metasurface with five supercells along the x-direction, whereas periodic boundary is assigned to the walls along y-direction to mimic an infinite array. Figure 6.16 depicts the 2D contour of scattering power intensity of the PB metasurface in both working states under the y-polarized incident EM wave. It is learned that the metasurface can split the incident LP beam into an LCP and an RCP beam traveling along two distinct directions in both "On" and "Off" states. However, the operation band identified from completely suppressed normal reflections and equal amplitudes of LCP and RCP beam dynamically switched from 8.1–12.2 GHz (a fractional bandwidth of 40.4%) in the "On" state to dual bands of 5.7–6.05 GHz and 9.15–10.5 GHz in the "Off" state. Outside these working bands, significant specular reflections appear which lower the PSHE efficiencies. The switched functionality of our PSHE device can be further inspected from Figs. 6.16c and 6.16f where the far-field scattering patterns for the "On" and "Off" state are plotted at a frequency around 8 GHz. In the former case, the incident LP beam was split and deflected to near $\pm 30.8°$, while in the latter case it was specularly reflected.

Though normal reflections were suppressed to some extent within 10.6–11.4 GHz in the "Off" state, the asymmetric LCP and RCP amplitudes and hybrid beams with degraded polarization extinct ratio deteriorate the PSHE efficiency even below 50%. In both cases, the simulated steering angles coincide well with those predicted from generalized Snell's law. The slight weak power intensity around 10.6 GHz in both "On" and "Off" states is attributed to the absorption loss of parasitic resonances illustrated in Fig. 6.15.

To further quantitatively study the PSHE performance, Fig. 6.17 shows the scattered intensity curve vs. the elevation angle. To illustrate the advantage of our design, we also show here the results of PB metasurface without chip inductors introduced in biasing lines for reference. As is shown, very symmetric scattering patterns with equal amplitude are inspected for our designed PB metasurface in both states. However, asymmetric patterns with unequal amplitude are observed for reference metasurface. The flowing current in the bottom plane gives rise to the asymmetric LCP and RCP patterns; see Fig. 6.18. The current density decreases significantly on both ring and feeding line when chip inductors are introduced. Therefore, the leakage of the RF signal through the bias line and ring is considerably prevented using chip inductors. The PSHE efficiency, calculated as the ratio between anomalously reflected power and the totally reflected power calculated by integrating power over the angle regions spanned by reflection modes, is obtained more than 89% at 5.95 GHz in the "On" state and 93% in the "Off" state within the operation band revealed in Fig. 6.16. At most frequency studied, the PSHE efficiency is achieved almost 100% such as at 8.7 and 10.8 GHz shown in Figs. 6.17c,d whose efficiency is obtained as 98%. The possibility of completely suppressing one polarization while maximizing the other can be implemented by loading asymmetric resistance along the gradient (x) direction.

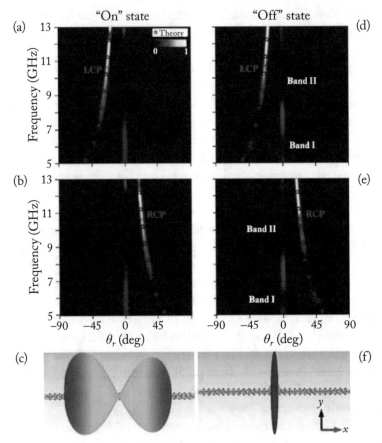

Figure 6.16: Simulated and theoretically calculated (symbols) normalized far-field scattering power intensity $P(\theta_r, \lambda)$ of the PB metasurface under y-polarized normal incidence for $-90° < \theta_r < 90°$. The $P(\theta_r, \lambda)$ is normalized to the maximum intensity. The (a, d) LCP and (b, e) RCP wave for (a–c) "On" state and (d–f) "Off" state. The 3D scattering pattern at 8 GHz in the (c) "On" state and (f) "Off" state.

This strategy has been validated from the results shown in Figs. 6.17g and 6.17h were almost completely suppressed RCP wave was inspected by setting $R_s = 200\ \Omega$ in the last three PIN diodes along $+x$ direction while $R_s = 2\ \Omega$ in all residual ones.

6.4   SUMMARY

This Chapter summarized our efforts on realizing tunable metadevices to dynamically control the properties of CP EM waves, mainly in the microwave regime where PIN diodes can be in-

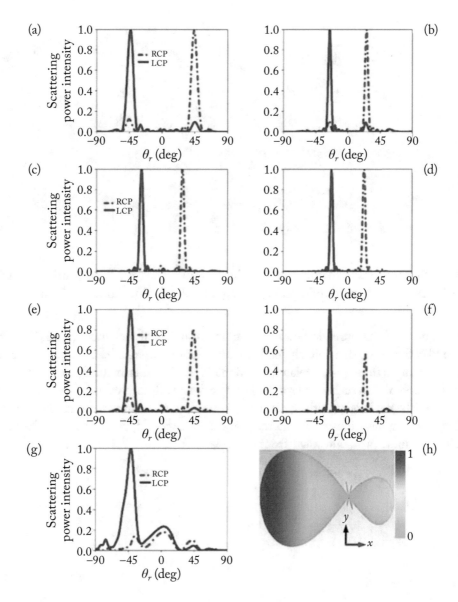

Figure 6.17: Scattering power intensity of the tunable PB metasurface (a–d) with (e, f) and without chip inductors in biasing lines at (a) 5.95, (b) 9.6, (e) 6, and (f) 10 GHz in the "Off" state, and at (c) 8.7 and (d) 10.8 GHz in the "On" state. Scattering (g) power intensity and (h) far-field pattern of the tunable PB metasurface at 8 GHz in the "On" state with $R_s = 200\ \Omega$ in the last three PIN diodes along +x direction and $R_s = 2\ \Omega$ in all residual ones.

(a)   (b)

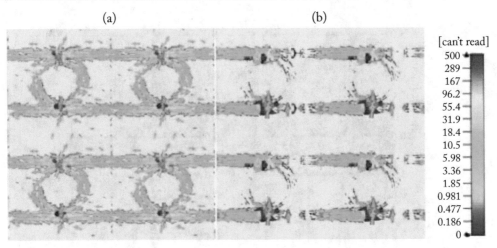

Figure 6.18: Surface current distributions on the bottom plane for PB metasurface (a) without and (b) with chip inductors introduced in feeding lines. Each plot shows 2 × 2 meta-atoms.

corporated into the meta-atom designs. The working principle of such metadevices is: switching on/off the PIN diode can dynamically modify the reflection phases of each meta-atom and thus significantly changes those phase-related functionalities. We experimentally demonstrated several metadevices based on such a scheme, including a helicity dynamical modulator for CP waves and a tunable PB metadevice with dynamically controlled CP-wave-manipulation functionalities. Many more applications of such a scheme can be expected, especially in high-frequency regimes and for more profound wave-front controls.

CHAPTER 7

Conclusions and Perspectives

This book presented an overview of our efforts in exploring the physics, designs, numerical and experimental demonstrations of multifunctional metasurfaces for achieving broadband and high-efficiency controls on EM waves with both linear and circular polarization, based on both passive and active schemes. Here, we conclude this book by mentioning the key challenges in this field and some promising future directions.

In reviewing the development of this field, we find that the meta-atom design is of importance since any new type of high-performance multifunctional meta-atom can surely stimulate a series of metadevices with diversified functionalities. We already see many interesting new designs in this field (for example, the polarization-dependent full-space meta-atom) which help people realize useful functional devices with more flexibilities in controlling EM waves. We also find that the materials in different frequency ranges have different behavior, the geometry of the meta-atoms required to achieve a chosen functionality is necessarily different for microwave, IR, or optical frequencies. For example, in the microwave regime metals behave as perfect electric conductors, the metallic losses can be very significant at IR and optical frequencies. As a result, while meta-atoms with multilayer metallic structures are widely used in designing high-efficiency transmissive multifunctional metasurfaces in low-frequency domains, these meta-atoms structures are difficult to be used directly at high frequencies because of the fabrication challenges and the material losses. Instead, simple metallic structures (say, metal bars) are currently used in designing transmission-mode multifunctional metadevices at high frequencies, although the realized metadevices suffer from low-efficiency issues. Very recently, all-dielectric metasurfaces began to appear, which seems to be a very promising route to overcome such an issue for transmissive multifunctional metadevices at high frequencies [76, 108, 119, 146, 228]. Along with the fast development in nanofabrication technologies, we expect that more excellent works can appear in this field, generating metadevices that can be eventually used in practice.

In addition to meta-atom design, we find that choosing the most powerful tuning scheme in the frequency regime of interest is very important. While pin/varactor dipoles work very well in the GHz regime for constructing tunable metadevices, they do not work well at frequencies higher than GHz. One commonly used tuning approach is to modulate the free carrier densities in materials (including semiconductors, 2D materials, and transparent conducting oxides) through electric gating or photoexcitation, which can work in the frequency range from THz to visible. Tuning material properties through certain phase transitions are also widely used in constructing active metasurfaces, which can work in the spectral range from GHz to visible.

Mechanical tuning offers another effective way to switch the EM properties of metadevices by reconfiguring the shape and surrounding environment of meta-atoms, which typically work in the THz regime.

Before concluding this book, we would like to mention several important future directions in the field of the multifunctional metasurface, based on our perspectives.

1. **Multifunctional metasurface with more functions**. For example, we can achieve the multifunctional metasurface based on the two-faced spatial asymmetric wavefront manipulation, i.e., one functionality for a wave propagating along one direction but a decoupled different one for the opposite direction [229, 230]. Also, we can implement multiple functions simultaneously at different frequencies.

2. **Spatiotemporal tunable metasurfaces.** Recently, as a new branch of active metasurface research, spatiotemporal metasurfaces, incorporating both spatial- and time-varying gradients of abrupt phases, have attracted rapidly growing interests, due to the stimulated new physical effects not presented in their static counterparts. For example, the time-varying gradient can impart additional frequency to the incoming EM wave, thereby resulting in a Doppler-like shift in the frequency. Shalaev and coworkers showed that spatiotemporal metasurfaces can lead to a generalized Snell's law, where both momentum and energy conservation are relaxed and several fascinating applications could be achieved [231]. Cui et al. experimentally demonstrated that the harmonics of the scattering wave can be freely manipulated in both amplitude and phase with time-varying metasurfaces, by controlling the time-sequences of the external sources [232, 233]. We expect more fascinating discoveries to appear in this subfield.

3. **Active metadevices for controlling waves other than EM wave.** Since the concept of metasurfaces/metamaterials can be extended to manipulate waves other than EM wave (e.g., acoustic wave [234–237], water wave [238], and the thermal wave [239, 240]), many research efforts have been devoted in the new subfield of tunable acoustic and thematic metasurfaces, leading to many fancy physical effect and practical applications. For example, Li et al. theoretically and experimentally demonstrated that robust and switchable acoustic asymmetric transmission can be achieved through gradient-index metasurfaces by harnessing carefully tailored losses [235]. Huang and coworkers established an approach to design switchable thermal cloaking and experimentally realized a macroscopic thermal diode of huge potential applications related to heat preservation and dissipation [239].

4. **Tunable metasystem for real applications.** Facing a different application scenario, different tunable and reconfigurable metadevices can be implemented based on a different structure with different tuning mechanisms and active materials. However, for real applications, lots of efforts are still needed, for example, in integrating tunable metasurfaces with conventional electro-optical devices, in large area low-cost fabrications, and in constructing software-defined platforms for automatic control on metadevices.

References

[1] R. A. Shelby, D. R. Smith, and S. Schultz, Experimental verification of a negative index of refraction, *Science*, 292:77–79, 2001. DOI: 10.1126/science.1058847. 1

[2] V. G. Veselago, The electrodynamics of substances with simultaneously negative values of ϵ and μ, *Soviet Physics Uspekhi*, 10:509–514, 1968. DOI: 10.1070/pu1968v010n04abeh003699. 1

[3] J. B. Pendry, Negative refraction makes a perfect lens, *Phys. Rev. Lett.*, 85:3966, 2000. DOI: 10.1103/physrevlett.85.3966. 1

[4] J. Valentine, S. Zhang, T. Zentgraf, E. Ulin-Avila, D. A. Genov, G. Bartal, and X. Zhang, Three-dimensional optical metamaterial with a negative refractive index, *Nature*, 455:376–379, 2008. DOI: 10.1038/nature07247. 1

[5] Z. Liu, H. Lee, Y. Xiong, C. Sun, and X. Zhang, Far-field optical hyperlens magnifying sub-diffraction-limited objects, *Science*, 315:1686, 2007. DOI: 10.1126/science.1137368. 1

[6] N. Fang, H. Lee, C. Sun, and X. Zhang, Sub–diffraction-limited optical imaging with a silver superlens, *Science*, 308:534–537, 2005. DOI: 10.1126/science.1108759. 1

[7] J. B. Pendry, D. Schurig, and D. R. Smith, Controlling electromagnetic fields, *Science*, 312:1780–1782, 2006. DOI: 10.1126/science.1125907. 1

[8] R. Liu, C. Ji, J. J. Mock, J. Y. Chin, T. J. Cui, and D. R. Smith, Broadband ground-plane cloak, *Science*, 323:366–369, 2009. DOI: 10.1126/science.1166949. 1

[9] H. F. Ma and T. J. Cui, Three-dimensional broadband ground-plane cloak made of metamaterials, *Nat. Commun.*, 1:21, 2010. DOI: 10.1038/ncomms1023. 1

[10] W. Sun, Q. He, J. Hao, and L. Zhou, A transparent metamaterial to manipulate electromagnetic wave polarizations, *Opt. Lett.*, 36:927–929, 2011. DOI: 10.1364/ol.36.000927. 1, 72

[11] S. Ma, X. Wang, W. Luo, S. Sun, Y. Zhang, Q. He, and L. Zhou, Ultra-wide band reflective metamaterial wave plates for terahertz waves, *EPL (Europhysics Letters)*, 117:37007, 2017. DOI: 10.1209/0295-5075/117/37007. 1

[12] J. Hao, Q. Ren, Z. An, X. Huang, Z. Chen, M. Qiu, and L. Zhou, Optical meta-material for polarization control, *Phys. Rev. A*, 80:23807, 2009. DOI: 10.1103/phys-reva.80.023807. 1, 63

[13] J. Hao, Y. Yuan, L. Ran, T. Jiang, J. A. Kong, C. T. Chan, and L. Zhou, Manipu-lating electromagnetic wave polarizations by anisotropic metamaterials, *Phys. Rev. Lett.*, 99:63908, 2007. DOI: 10.1103/physrevlett.99.063908. 1, 40, 63, 123

[14] J. Hao, J. Wang, X. Liu, W. J. Padilla, L. Zhou, and M. Qiu, High performance optical absorber based on a plasmonic metamaterial, *Appl. Phys. Lett.*, 96:251104, 2010. DOI: 10.1063/1.3442904. 1

[15] Z. Song, Q. He, S. Xiao, and L. Zhou, Making a continuous metal film transparent via scattering cancellations, *Appl. Phys. Lett.*, 101:181110, 2012. DOI: 10.1063/1.4764945. 1

[16] R. Malureanu, M. Zalkovskij, Z. Song, C. Gritti, A. Andryieuski, Q. He, L. Zhou, P. U. Jepsen, and A. V. Lavrinenko, A new method for obtaining transparent electrodes, *Opt. Express*, 20:22770–22782, 2012. DOI: 10.1364/oe.20.022770. 1

[17] S. Enoch, G. Tayeb, P. Sabouroux, N. Guérin, and P. Vincent, A metamaterial for direc-tive emission, *Phys. Rev. Lett.*, 89:213902, 2002. DOI: 10.1103/physrevlett.89.213902. 1

[18] Z. H. Jiang, Q. Wu, and D. H. Werner, Demonstration of enhanced broadband unidi-rectional electromagnetic radiation enabled by a subwavelength profile leaky anisotropic zero-index metamaterial coating, *Phys. Rev. B*, 86:125131, 2012. DOI: 10.1103/phys-revb.86.125131. 1

[19] S. Jeon, E. Menard, J. U. Park, J. Maria, M. Meitl, J. Zaumseil, and J. A. Rogers, Three-dimensional nanofabrication with rubber stamps and conformable photomasks, *Adv. Mater.*, 16:1369–1373, 2004. DOI: 10.1002/adma.200400593. 2

[20] M. S. Rill, C. Plet, M. Thiel, I. Staude, G. Von Freymann, S. Linden, and M. Wegener, Photonic metamaterials by direct laser writing and silver chemical vapour deposition, *Nat. Mater.*, 7:543–546, 2008. DOI: 10.1038/nmat2197. 2

[21] X. Ni, N. K. Emani, A. V. Kildishev, A. Boltasseva, and V. M. Shalaev, Broadband light bending with plasmonic nanoantennas, *Science*, 335:427, 2012. DOI: 10.1126/sci-ence.1214686. 2, 27, 33, 43, 46, 61, 65, 75, 87

[22] A. Pors, O. Albrektsen, I. P. Radko, and S. I. Bozhevolnyi, Gap plasmon-based metasur-faces for total control of reflected light, *Sci. Rep.*, 3:2155, 2013. DOI: 10.1038/srep02155. 2

[23] S. Sun, K. Yang, C. Wang, T. Juan, W. T. Chen, C. Y. Liao, Q. He, S. Xiao, W. Kung, and G. Guo, High-efficiency broadband anomalous reflection by gradient meta-surfaces, *Nano Lett.*, 12:6223–6229, 2012. DOI: 10.1021/nl3032668. 2, 61, 65, 75, 87

[24] C. Pfeiffer and A. Grbic, Metamaterial Huygens' surfaces: Tailoring wave fronts with reflectionless sheets, *Phys. Rev. Lett.*, 110:197401, 2013. DOI: 10.1103/physrevlett.110.197401. 2, 61, 65, 72, 75, 83, 87

[25] J. Luo, H. Yu, M. Song, and Z. Zhang, Highly efficient wavefront manipulation in terahertz based on plasmonic gradient metasurfaces, *Opt. Lett.*, 39:2229–2231, 2014. DOI: 10.1364/ol.39.002229. 2

[26] C. Pfeiffer, N. K. Emani, A. M. Shaltout, A. Boltasseva, V. M. Shalaev, and A. Grbic, Efficient light bending with isotropic metamaterial Huygens' surfaces, *Nano Lett.*, 14:2491–2497, 2014. DOI: 10.1021/nl5001746. 2, 40

[27] Z. Wei, Y. Cao, X. Su, Z. Gong, Y. Long, and H. Li, Highly efficient beam steering with a transparent metasurface, *Opt. Express*, 21:10739–10745, 2013. DOI: 10.1364/oe.21.010739. 2

[28] N. Yu, P. Genevet, M. A. Kats, F. Aieta, J. Tetienne, F. Capasso, and Z. Gaburro, Light propagation with phase discontinuities: Generalized laws of reflection and refraction, *Science*, 334:333–337, 2011. DOI: 10.1126/science.1210713. 2, 27, 33, 43, 46, 61, 62, 65, 75, 87

[29] Z. Miao, Q. Wu, X. Li, Q. He, K. Ding, Z. An, Y. Zhang, and L. Zhou, Widely tunable terahertz phase modulation with gate-controlled graphene metasurfaces, *Phys. Rev. X*, 5:41027, 2015. DOI: 10.1103/physrevx.5.041027. 2

[30] S. Sun, Q. He, S. Xiao, Q. Xu, X. Li, and L. Zhou, Gradient-index meta-surfaces as a bridge linking propagating waves and surface waves, *Nat. Mater.*, 11:426–431, 2012. DOI: 10.1038/nmat3292. 2, 27, 33, 36, 43, 46, 61, 62, 65, 75, 87

[31] W. Sun, Q. He, S. Sun, and L. Zhou, High-efficiency surface plasmon meta-couplers: Concept and microwave-regime realizations, *Light Sci. Applic.*, 5:e16003, 2016. DOI: 10.1038/lsa.2016.3. 2, 72

[32] M. J. Lockyear, A. P. Hibbins, and J. R. Sambles, Microwave surface-plasmon-like modes on thin metamaterials, *Phys. Rev. Lett.*, 102:73901, 2009. DOI: 10.1103/physrevlett.102.073901. 2

[33] X. Xiong, Y. Hu, S. Jiang, Y. Hu, R. Fan, G. Ma, D. Shu, R. Peng, and M. Wang, Metallic stereostructured layer: An approach for broadband polarization state manipulation, *Appl. Phys. Lett.*, 105:201105, 2014. DOI: 10.1063/1.4902405. 2

[34] F. Ding, Z. Wang, S. He, V. M. Shalaev, and A. V. Kildishev, Broadband high-efficiency half-wave plate: A supercell-based plasmonic metasurface approach, *ACS Nano*, 9:4111–4119, 2015. DOI: 10.1021/acsnano.5b00218. 2, 40

[35] Y. Yang, W. Wang, P. Moitra, I. I. Kravchenko, D. P. Briggs, and J. Valentine, Dielectric meta-reflectarray for broadband linear polarization conversion and optical vortex generation, *Nano Lett.*, 14:1394–1399, 2014. DOI: 10.1021/nl4044482. 2

[36] S. Jiang, X. Xiong, Y. Hu, G. Ma, R. Peng, C. Sun, and M. Wang, Controlling the polarization state of light with a dispersion-free metastructure, *Phys. Rev. X*, 4:21026, 2014. DOI: 10.1103/physrevx.4.021026. 2, 70, 93

[37] C. Pfeiffer and A. Grbic, Bianisotropic metasurfaces for optimal polarization control: Analysis and synthesis, *Phys. Rev. Appl.*, 2:44011, 2014. DOI: 10.1103/physrevapplied.2.044011. 2

[38] C. Pfeiffer, C. Zhang, V. Ray, L. J. Guo, and A. Grbic, High performance bianisotropic metasurfaces: Asymmetric transmission of light, *Phys. Rev. Lett.*, 113:23902, 2014. DOI: 10.1103/physrevlett.113.023902. 2, 40

[39] J. Yun, S. Kim, H. Yun, K. Lee, J. Sung, J. Kim, Y. Lee, and B. Lee, Broadband ultrathin circular polarizer at visible and near-infrared wavelengths using a non-resonant characteristic in helically stacked nano-gratings, *Opt. Express*, 25:14260–14269, 2017. DOI: 10.1364/oe.25.014260. 2

[40] X. Li, S. Xiao, B. Cai, Q. He, T. J. Cui, and L. Zhou, Flat metasurfaces to focus electromagnetic waves in reflection geometry, *Opt. Lett.*, 37:4940–4942, 2012. DOI: 10.1364/ol.37.004940. 2

[41] F. Aieta, M. A. Kats, P. Genevet, and F. Capasso, Multiwavelength achromatic metasurfaces by dispersive phase compensation, *Science*, 347:1342–1345, 2015. DOI: 10.1126/science.aaa2494. 2

[42] X. Ma, M. Pu, X. Li, C. Huang, Y. Wang, W. Pan, B. Zhao, J. Cui, C. Wang, and Z. Zhao, A planar chiral meta-surface for optical vortex generation and focusing, *Sci. Rep.*, 5:10365, 2015. DOI: 10.1038/srep10365. 2

[43] L. Huang, X. Chen, H. Mühlenbernd, H. Zhang, S. Chen, B. Bai, Q. Tan, G. Jin, K. Cheah, and C. Qiu, Three-dimensional optical holography using a plasmonic metasurface, *Nat. Commun.*, 4:2808, 2013. DOI: 10.1038/ncomms3808. 2, 138

[44] G. Zheng, H. Mühlenbernd, M. Kenney, G. Li, T. Zentgraf, and S. Zhang, Metasurface holograms reaching 80% efficiency, *Nat. Nanotechnol.*, 10:308–312, 2015. DOI: 10.1038/nnano.2015.2. 2, 70

[45] G. Lee, G. Yoon, S. Lee, H. Yun, J. Cho, K. Lee, H. Kim, J. Rho, and B. Lee, Complete amplitude and phase control of light using broadband holographic metasurfaces, *Nanoscale*, 10:4237–4245, 2018. DOI: 10.1039/c7nr07154j. 2

[46] M. Khorasaninejad, W. T. Chen, R. C. Devlin, J. Oh, A. Y. Zhu, and F. Capasso, Metalenses at visible wavelengths: Diffraction-limited focusing and subwavelength resolution imaging, *Science*, 352:1190–1194, 2016. DOI: 10.1126/science.aaf6644. 2

[47] A. Arbabi, Y. Horie, A. J. Ball, M. Bagheri, and A. Faraon, Subwavelength-thick lenses with high numerical apertures and large efficiency based on high-contrast transmitarrays, *Nat. Commun.*, 6:7069, 2015. DOI: 10.1038/ncomms8069. 2

[48] X. Chen, M. Chen, M. Q. Mehmood, D. Wen, F. Yue, C. W. Qiu, and S. Zhang, Longitudinal multifoci metalens for circularly polarized light, *Adv. Opt. Mater.*, 3:1201–1206, 2015. DOI: 10.1002/adom.201500110. 2, 138

[49] C. Pfeiffer and A. Grbic, Cascaded metasurfaces for complete phase and polarization control, *Appl. Phys. Lett.*, 102:231116, 2013. DOI: 10.1063/1.4810873. 2

[50] J. S. Ho, B. Qiu, Y. Tanabe, A. J. Yeh, S. Fan, and A. S. Poon, Planar immersion lens with metasurfaces, *Phys. Rev. B*, 91:125145, 2015. DOI: 10.1103/physrevb.91.125145. 2

[51] J. Y. Kim, H. Kim, B. H. Kim, T. Chang, J. Lim, H. M. Jin, J. H. Mun, Y. J. Choi, K. Chung, and J. Shin, Highly tunable refractive index visible-light metasurface from block copolymer self-assembly, *Nat. Commun.*, 7:12911, 2016. DOI: 10.1038/ncomms12911. 2

[52] J. Park, J. Kang, S. J. Kim, X. Liu, and M. L. Brongersma, Dynamic reflection phase and polarization control in metasurfaces, *Nano Lett.*, 17:407–413, 2017. DOI: 10.1021/acs.nanolett.6b04378. 2

[53] Y. Shi and S. Fan, Dynamic non-reciprocal meta-surfaces with arbitrary phase reconfigurability based on photonic transition in meta-atoms, *Appl. Phys. Lett.*, 108:21110, 2016. DOI: 10.1063/1.4939915. 2

[54] W. Luo, S. Xiao, Q. He, S. Sun, and L. Zhou, Photonic spin Hall effect with nearly 100% efficiency, *Adv. Opt. Mater.*, 3:1102–1108, 2015. DOI: 10.1002/adom.201500068. 2, 7, 12, 40, 61, 63, 71, 78, 79

[55] X. Yin, Z. Ye, J. Rho, Y. Wang, and X. Zhang, Photonic spin Hall effect at metasurfaces, *Science*, 339:1405–1407, 2013. DOI: 10.1126/science.1231758. 2, 138

[56] A. Shaltout, J. Liu, A. Kildishev, and V. Shalaev, Photonic spin Hall effect in gap-plasmon metasurfaces for on-chip chiroptical spectroscopy, *Optica*, 2:860–863, 2015. DOI: 10.1364/optica.2.000860. 2, 70

[57] T. Cai, S. Tang, G. Wang, H. X. Xu, S. Sun, Q. He, and L. Zhou, High-performance bifunctional metasurfaces in transmission and reflection geometries, *Adv. Opt. Mater.*, 5:1600506, 2017. DOI: 10.1002/adom.201600506. 2, 3, 13, 48, 83

[58] T. Cai, G. Wang, S. Tang, H. X. Xu, J. Duan, H. Guo, F. Guan, S. Sun, Q. He, and L. Zhou, High-efficiency and full-space manipulation of electromagnetic wave fronts with metasurfaces, *Phys. Rev. Appl.*, 8:34033, 2017. DOI: 10.1103/physrevapplied.8.034033. 2, 13

[59] F. Ding, R. Deshpande, and S. I. Bozhevolnyi, Bifunctional gap-plasmon metasurfaces for visible light: polarization-controlled unidirectional surface plasmon excitation and beam steering at normal incidence, *Light Sci. Applic.*, 7:17178, 2018. DOI: 10.1038/lsa.2017.178. 2, 3

[60] S. Boroviks, R. A. Deshpande, N. A. Mortensen, and S. I. Bozhevolnyi, Multifunctional metamirror: Polarization splitting and focusing, *ACS Photon.*, 5:1648–1653, 2017. DOI: 10.1021/acsphotonics.7b01091. 2

[61] H. X. Xu, S. Tang, X. Ling, W. Luo, and L. Zhou, Flexible control of highly-directive emissions based on bifunctional metasurfaces with low polarization cross-talking, *Ann. Phys.-Berlin*, 529:1700045, 2017. DOI: 10.1002/andp.201700045. 2, 13, 14, 48

[62] H. X. Xu, S. Tang, G. Wang, T. Cai, W. Huang, Q. He, S. Sun, and L. Zhou, Multi-functional microstrip array combining a linear polarizer and focusing metasurface, *IEEE T. Antenn. Propag.*, 64:3676–3682, 2016. DOI: 10.1109/tap.2016.2565742. 2

[63] S. Liu, T. J. Cui, A. Noor, Z. Tao, H. C. Zhang, G. D. Bai, Y. Yang, and X. Y. Zhou, Negative reflection and negative surface wave conversion from obliquely incident electromagnetic waves, *Light Sci. Applic.*, 7:18008, 2018. DOI: 10.1038/lsa.2018.8. 2

[64] W. T. Chen, K. Yang, C. Wang, Y. Huang, G. Sun, I. Chiang, C. Y. Liao, W. Hsu, H. T. Lin, and S. Sun, High-efficiency broadband meta-hologram with polarization-controlled dual images, *Nano Lett.*, 14:225–230, 2014. DOI: 10.1021/nl403811d. 3, 9

[65] A. Arbabi, Y. Horie, M. Bagheri, and A. Faraon, Dielectric metasurfaces for complete control of phase and polarization with subwavelength spatial resolution and high transmission, *Nat. Nanotechnol.*, 10:937, 2015. DOI: 10.1038/nnano.2015.186. 3

[66] L. Huang, H. Mühlenbernd, X. Li, X. Song, B. Bai, Y. Wang, and T. Zentgraf, Broadband hybrid holographic multiplexing with geometric metasurfaces, *Adv. Mater.*, 27:6444–6449, 2015. DOI: 10.1002/adma.201502541. 3, 10

[67] E. Maguid, I. Yulevich, D. Veksler, V. Kleiner, M. L. Brongersma, and E. Hasman, Photonic spin-controlled multifunctional shared-aperture antenna array, *Science*, 352:1202–1206, 2016. DOI: 10.1126/science.aaf3417. 3

[68] J. B. Mueller, N. A. Rubin, R. C. Devlin, B. Groever, and F. Capasso, Metasurface polarization optics: Independent phase control of arbitrary orthogonal states of polarization, *Phys. Rev. Lett.*, 118:113901, 2017. DOI: 10.1103/physrevlett.118.113901. 3

[69] S. Xiao, F. Zhong, H. Liu, S. Zhu, and J. Li, Flexible coherent control of plasmonic spin-Hall effect, *Nat. Commun.*, 6:8360, 2015. DOI: 10.1038/ncomms9360. 3

[70] S. Choudhury, U. Guler, A. Shaltout, V. M. Shalaev, A. V. Kildishev, and A. Boltasseva, Pancharatnam-Berry phase manipulating metasurface for visible color hologram based on low loss silver thin film, *Adv. Opt. Mater.*, 5:1700196, 2017. DOI: 10.1002/adom.201700196. 3

[71] D. Wen, S. Chen, F. Yue, K. Chan, M. Chen, M. Ardron, K. F. Li, P. W. H. Wong, K. W. Cheah, and E. Y. B. Pun, Metasurface device with helicity-dependent functionality, *Adv. Opt. Mater.*, 4:321–327, 2016. DOI: 10.1002/adom.201500498. 3, 10

[72] S. Wang, X. Wang, Q. Kan, J. Ye, S. Feng, W. Sun, P. Han, S. QU, and Y. Zhang, Spin-selected focusing and imaging based on metasurface lens, *Opt. Express*, 23:26434–26441, 2015. DOI: 10.1364/OE.23.026434. 3

[73] E. Maguid, I. Yulevich, M. Yannai, V. Kleiner, M. L. Brongersma, and E. Hasman, Multifunctional interleaved geometric-phase dielectric metasurfaces, *Light Sci. Applic.*, 6:e17027, 2017. DOI: 10.1038/lsa.2017.27. 3

[74] D. Veksler, E. Maguid, N. Shitrit, D. Ozeri, V. Kleiner, and E. Hasman, Multiple wavefront shaping by metasurface based on mixed random antenna groups, *ACS Photon.*, 2:661–667, 2015. DOI: 10.1021/acsphotonics.5b00113. 3

[75] X. Li, L. Chen, Y. Li, X. Zhang, M. Pu, Z. Zhao, X. Ma, Y. Wang, M. Hong, and X. Luo, Multicolor 3D meta-holography by broadband plasmonic modulation, *Sci. Adv.*, 2:e1601102, 2016. DOI: 10.1126/sciadv.1601102. 3

[76] B. Wang, F. Dong, Q. Li, D. Yang, C. Sun, J. Chen, Z. Song, L. Xu, W. Chu, and Y. Xiao, Visible-frequency dielectric metasurfaces for multiwavelength achromatic and highly dispersive holograms, *Nano Lett.*, 16:5235–5240, 2016. DOI: 10.1021/acs.nanolett.6b02326. 3, 12, 151

[77] S. M. Kamali, E. Arbabi, A. Arbabi, Y. Horie, M. Faraji-Dana, and A. Faraon, Angle-multiplexed metasurfaces: Encoding independent wavefronts in a single metasurface under different illumination angles, *Phys. Rev. X*, 7:41056, 2017. DOI: 10.1103/physrevx.7.041056. 3

[78] M. Qiu, M. Jia, S. Ma, S. Sun, Q. He, and L. Zhou, Angular dispersions in tera-hertz metasurfaces: Physics and applications, *Phys. Rev. Appl.*, 9:54050, 2018. DOI: 10.1103/physrevapplied.9.054050. 3

[79] L. Zhou, Controlling angle dispersions in optical metasurfaces, *Light Sci. Applic.*, 2020. DOI: 10.1038/s41377-020-0313-0. 4

[80] B. Zhu, Y. Feng, J. Zhao, C. Huang, and T. Jiang, Switchable metamaterial reflec-tor/absorber for different polarized electromagnetic waves, *Appl. Phys. Lett.*, 97:51906, 2010. DOI: 10.1063/1.3477960. 4

[81] H. X. Xu, S. Sun, S. Tang, S. Ma, Q. He, G. Wang, T. Cai, H. Li, and L. Zhou, Dy-namical control on helicity of electromagnetic waves by tunable metasurfaces, *Sci. Rep.*, 6:27503, 2016. DOI: 10.1038/srep27503. 4, 123

[82] Z. Tao, X. Wan, B. C. Pan, and T. J. Cui, Reconfigurable conversions of reflection, trans-mission, and polarization states using active metasurface, *Appl. Phys. Lett.*, 110:121901, 2017. DOI: 10.1063/1.4979033. 4

[83] J. Zhao, Q. Cheng, J. Chen, M. Q. Qi, W. X. Jiang, and T. J. Cui, A tunable metamate-rial absorber using varactor diodes, *New J. Phys.*, 15:43049, 2013. DOI: 10.1088/1367-2630/15/4/043049. 4

[84] W. Xu and S. Sonkle, Microwave diode switchable metamaterial reflector/absorber, *Appl. Phys. Lett.*, 103:31902, 2013. DOI: 10.1063/1.4813750. 4

[85] H. Chen, W. J. Padilla, J. M. Zide, A. C. Gossard, A. J. Taylor, and R. D. Averitt, Active terahertz metamaterial devices, *Nature*, 444:597–600, 2006. DOI: 10.1038/nature05343. 4, 93, 121

[86] H. Chen, W. J. Padilla, M. J. Cich, A. K. Azad, R. D. Averitt, and A. J. Taylor, A metamaterial solid-state terahertz phase modulator, *Nat. Photon.*, 3:148–151, 2009. DOI: 10.1038/nphoton.2009.3. 4

[87] A. Y. Zhu, A. I. Kuznetsov, B. Luk Yanchuk, N. Engheta, and P. Genevet, Traditional and emerging materials for optical metasurfaces, *Nanophotonics*, 6:452–471, 2017. DOI: 10.1515/nanoph-2016-0032. 4

[88] A. M. Shaltout, N. Kinsey, J. Kim, R. Chandrasekar, J. C. Ndukaife, A. Boltasseva, and V. M. Shalaev, Development of optical metasurfaces: Emerging concepts and new materials, *P. IEEE*, 104:2270–2287, 2016. DOI: 10.1109/jproc.2016.2590882. 4

[89] D. Traviss, R. Bruck, B. Mills, M. Abb, and O. L. Muskens, Ultrafast plasmonics using transparent conductive oxide hybrids in the epsilon-near-zero regime, *Appl. Phys. Lett.*, 102:121112, 2013. DOI: 10.1063/1.4798833. 4

[90] A. Forouzmand, M. M. Salary, S. Inampudi, and H. Mosallaei, A tunable multigate indium-tin-oxide-assisted all-dielectric metasurface, *Adv. Opt. Mater.*, 6:1701275, 2018. DOI: 10.1002/adom.201701275. 4

[91] G. V. Naik, J. Kim, and A. Boltasseva, Oxides and nitrides as alternative plasmonic materials in the optical range, *Opt. Mater. Express*, 1:1090–1099, 2011. DOI: 10.1364/ome.1.001090. 4

[92] F. Bonaccorso, Z. Sun, T. Hasan, and A. Ferrari, Graphene photonics and optoelectronics, *Nat. Photonics*, 4:611, 2010. DOI: 10.1038/nphoton.2010.186. 4

[93] F. H. Koppens, D. E. Chang, and F. J. Garcia de Abajo, Graphene plasmonics: A platform for strong light-matter interactions, *Nano Lett.*, 11:3370–3377, 2011. DOI: 10.1021/nl201771h. 4

[94] L. Ju, B. Geng, J. Horng, C. Girit, M. Martin, Z. Hao, H. A. Bechtel, X. Liang, A. Zettl, and Y. R. Shen, Graphene plasmonics for tunable terahertz metamaterials, *Nat. Nanotechnol.*, 6:630, 2011. DOI: 10.1038/nnano.2011.146. 4

[95] S. H. Lee, M. Choi, T. Kim, S. Lee, M. Liu, X. Yin, H. K. Choi, S. S. Lee, C. Choi, and S. Choi, Switching terahertz waves with gate-controlled active graphene metamaterials, *Nat. Mater.*, 11:936–941, 2012. DOI: 10.1038/nmat3433. 4

[96] N. Mou, S. Sun, H. Dong, S. Dong, Q. He, L. Zhou, and L. Zhang, Hybridization-induced broadband terahertz wave absorption with graphene metasurfaces, *Opt. Express*, 26:11728–11736, 2018. DOI: 10.1364/oe.26.011728. 4

[97] S. Shi, B. Zeng, H. Han, X. Hong, H. Tsai, H. Jung, A. Zettl, M. Crommie, and F. Wang, Optimizing broadband terahertz modulation with hybrid graphene/metasurface structures, *Nano Lett.*, 15:372–377, 2015. DOI: 10.1021/nl503670d. 4

[98] T. T. Kim, H. Kim, M. Kenney, H. S. Park, H. D. Kim, B. Min, and S. Zhang, Amplitude modulation of anomalously refracted terahertz waves with gated-graphene metasurfaces, *Adv. Opt. Mater.*, 6:1700507, 2018. DOI: 10.1002/adom.201700507. 4

[99] Y. Fan, N. Shen, T. Koschny, and C. M. Soukoulis, Tunable terahertz meta-surface with graphene cut-wires, *ACS Photon.*, 2:151–156, 2015. DOI: 10.1021/ph500366z. 4

[100] T. Kim, S. S. Oh, H. Kim, H. S. Park, O. Hess, B. Min, and S. Zhang, Electrical access to critical coupling of circularly polarized waves in graphene chiral metamaterials, *Sci. Adv.*, 3:e1701377, 2017. DOI: 10.1126/sciadv.1701377. 4

[101] B. Zeng, Z. Huang, A. Singh, Y. Yao, A. K. Azad, A. D. Mohite, A. J. Taylor, D. R. Smith, and H. Chen, Hybrid graphene metasurfaces for high-speed mid-infrared light modulation and single-pixel imaging, *Light Sci. Applic.*, 7:51, 2018. DOI: 10.1038/s41377-018-0055-4. 4

[102] Y. Yao, M. A. Kats, R. Shankar, Y. Song, J. Kong, M. Loncar, and F. Capasso, Wide wavelength tuning of optical antennas on graphene with nanosecond response time, *Nano Lett.*, 14:214–219, 2014. DOI: 10.1021/nl403751p. 4

[103] V. W. Brar, M. S. Jang, M. Sherrott, J. J. Lopez, and H. A. Atwater, Highly confined tunable mid-infrared plasmonics in graphene nanoresonators, *Nano Lett.*, 13:2541–2547, 2013. DOI: 10.1021/nl400601c. 4

[104] M. C. Sherrott, P. W. Hon, K. T. Fountaine, J. C. Garcia, S. M. Ponti, V. W. Brar, L. A. Sweatlock, and H. A. Atwater, Experimental demonstration of > 230 phase modulation in gate-tunable graphene—gold reconfigurable mid-infrared metasurfaces, *Nano Lett.*, 17:3027–3034, 2017. DOI: 10.1021/acs.nanolett.7b00359. 4

[105] N. Dabidian, S. Dutta-Gupta, I. Kholmanov, K. Lai, F. Lu, J. Lee, M. Jin, S. Trendafilov, A. Khanikaev, and B. Fallahazad, Experimental demonstration of phase modulation and motion sensing using graphene-integrated metasurfaces, *Nano Lett.*, 16:3607–3615, 2016. DOI: 10.1021/acs.nanolett.6b00732. 4

[106] Y. Yao, R. Shankar, M. A. Kats, Y. Song, J. Kong, M. Loncar, and F. Capasso, Electrically tunable metasurface perfect absorbers for ultrathin mid-infrared optical modulators, *Nano Lett.*, 14:6526–6532, 2014. DOI: 10.1021/nl503104n. 4

[107] Z. Li, K. Yao, F. Xia, S. Shen, J. Tian, and Y. Liu, Graphene plasmonic metasurfaces to steer infrared light, *Sci. Rep.*, 5:12423, 2015. DOI: 10.1038/srep12423. 4

[108] J. Sautter, I. Staude, M. Decker, E. Rusak, D. N. Neshev, I. Brener, and Y. S. Kivshar, Active tuning of all-dielectric metasurfaces, *ACS Nano*, 9:4308–4315, 2015. DOI: 10.1021/acsnano.5b00723. 4, 151

[109] D. Shrekenhamer, W. Chen, and W. J. Padilla, Liquid crystal tunable metamaterial absorber, *Phys. Rev. Lett.*, 110:177403, 2013. DOI: 10.1103/physrevlett.110.177403. 4

[110] J. Bohn, T. Bucher, K. E. Chong, A. Komar, D. Choi, D. N. Neshev, Y. S. Kivshar, T. Pertsch, and I. Staude, Active tuning of spontaneous emission by Mie-resonant dielectric metasurfaces, *Nano Lett.*, 18:3461–3465, 2018. DOI: 10.1021/acs.nanolett.8b00475. 4

[111] H. Chen, J. F. O'Hara, A. K. Azad, A. J. Taylor, R. D. Averitt, D. B. Shrekenhamer, and W. J. Padilla, Experimental demonstration of frequency-agile terahertz metamaterials, *Nat. Photon.*, 2:295, 2008. DOI: 10.1038/nphoton.2008.52. 4

[112] J. Gu, R. Singh, X. Liu, X. Zhang, Y. Ma, S. Zhang, S. A. Maier, Z. Tian, A. K. Azad, and H. Chen, Active control of electromagnetically induced transparency analogue in terahertz metamaterials, *Nat. Commun.*, 3:1151, 2012. DOI: 10.1038/ncomms2153. 4

[113] M. Ren, W. Wu, W. Cai, B. Pi, X. Zhang, and J. Xu, Reconfigurable metasurfaces that enable light polarization control by light, *Light Sci. Applic.*, 6:e16254, 2017. DOI: 10.1038/lsa.2016.254. 4

[114] M. Rudé, V. Mkhitaryan, A. E. Cetin, T. A. Miller, A. Carrilero, S. Wall, F. J. G. de Abajo, H. Altug, and V. Pruneri, Ultrafast and broadband tuning of resonant optical nanostructures using phase-change materials, *Adv. Opt. Mater.*, 4:1060–1066, 2016. DOI: 10.1002/adom.201600079. 4

[115] B. Gholipour, J. Zhang, K. F. MacDonald, D. W. Hewak, and N. I. Zheludev, An all-optical, non-volatile, bidirectional, phase-change meta-switch, *Adv. Mater.*, 25:3050–3054, 2013. DOI: 10.1002/adma.201300588. 4

[116] J. Rensberg, S. Zhang, Y. Zhou, A. S. Mcleod, C. Schwarz, M. Goldflam, M. Liu, J. Kerbusch, R. Nawrodt, and S. Ramanathan, Active optical metasurfaces based on defect-engineered phase-transition materials, *Nano Lett.*, 16:1050–1055, 2016. DOI: 10.1021/acs.nanolett.5b04122. 4

[117] X. Yin, T. Steinle, L. Huang, T. Taubner, M. Wuttig, T. Zentgraf, and H. Giessen, Beam switching and bifocal zoom lensing using active plasmonic metasurfaces, *Light Sci. Applic.*, 6:e17016, 2017. DOI: 10.1038/lsa.2017.16. 4

[118] J. Ou, E. Plum, J. Zhang, and N. I. Zheludev, An electromechanically reconfigurable plasmonic metamaterial operating in the near-infrared, *Nat. Nanotechnol.*, 8:252–255, 2013. DOI: 10.1038/nnano.2013.25. 4

[119] S. Sun, W. Yang, C. Zhang, J. Jing, Y. Gao, X. Yu, Q. Song, and S. Xiao, Real-time tunable colors from microfluidic reconfigurable all-dielectric metasurfaces, *ACS Nano*, 12:2151–2159, 2018. DOI: 10.1021/acsnano.7b07121. 4, 151

[120] N. I. Zheludev and E. Plum, Reconfigurable nanomechanical photonic metamaterials, *Nat. Nanotechnol.*, 11:16, 2016. DOI: 10.1038/nnano.2015.302. 4

[121] Y. H. Fu, A. Q. Liu, W. M. Zhu, X. M. Zhang, D. P. Tsai, J. B. Zhang, T. Mei, J. F. Tao, H. C. Guo, and X. H. Zhang, A micromachined reconfigurable metamaterial via reconfiguration of asymmetric split-ring resonators, *Adv. Funct. Mater.*, 21:3589–3594, 2011. DOI: 10.1002/adfm.201101087. 4

[122] H. Tao, A. Strikwerda, K. Fan, W. Padilla, X. Zhang, and R. Averitt, Reconfigurable terahertz metamaterials, *Phys. Rev. Lett.*, 103:147401, 2009. DOI: 10.1103/physrevlett.103.147401. 4

[123] L. Cong, P. Pitchappa, Y. Wu, L. Ke, C. Lee, N. Singh, H. Yang, and R. Singh, Active multifunctional microelectromechanical system metadevices: Applications in polarization

control, wavefront deflection, and holograms, *Adv. Opt. Mater.*, 5:1600716, 2017. DOI: 10.1002/adom.201600716. 4

[124] L. Cong, P. Pitchappa, C. Lee, and R. Singh, Active phase transition via loss engineering in a terahertz MEMS metamaterial, *Adv. Mater.*, 29:1700733, 2017. DOI: 10.1002/adma.201700733. 4

[125] P. Gutruf, C. Zou, W. Withayachumnankul, M. Bhaskaran, S. Sriram, and C. Fumeaux, Mechanically tunable dielectric resonator metasurfaces at visible frequencies, *ACS Nano*, 10:133–141, 2016. DOI: 10.1021/acsnano.5b05954. 5

[126] L. Zhu, J. Kapraun, J. Ferrara, and C. J. Chang-Hasnain, Flexible photonic metastructures for tunable coloration, *Optica*, 2:255–258, 2015. DOI: 10.1364/optica.2.000255. 5

[127] S. C. Malek, H. Ee, and R. Agarwal, Strain multiplexed metasurface holograms on a stretchable substrate, *Nano Lett.*, 17:3641–3645, 2017. DOI: 10.1021/acs.nanolett.7b00807. 5

[128] S. M. Kamali, E. Arbabi, A. Arbabi, Y. Horie, and A. Faraon, Highly tunable elastic dielectric metasurface lenses, *Laser Photon. Rev.*, 10:1002–1008, 2016. DOI: 10.1002/lpor.201600144. 5

[129] T. Driscoll, H. Kim, B. Chae, B. Kim, Y. Lee, N. M. Jokerst, S. Palit, D. R. Smith, M. Di Ventra, and D. N. Basov, Memory metamaterials, *Science*, 325:1518–1521, 2009. DOI: 10.1126/science.1176580. 5

[130] D. Wang, L. Zhang, Y. Gu, M. Mehmood, Y. Gong, A. Srivastava, L. Jian, T. Venkatesan, C. Qiu, and M. Hong, Switchable ultrathin quarter-wave plate in terahertz using active phase-change metasurface, *Sci. Rep.*, 5:15020, 2015. DOI: 10.1038/srep15020. 5

[131] G. Scalari, C. Maissen, S. Cibella, R. Leoni, and J. Faist, High quality factor, fully switchable terahertz superconducting metasurface, *Appl. Phys. Lett.*, 105:261104, 2014. DOI: 10.1063/1.4905199. 5

[132] B. Jin, C. Zhang, S. Engelbrecht, A. Pimenov, J. WU, Q. Xu, C. Cao, J. Chen, W. Xu, and L. Kang, Low loss and magnetic field-tunable superconducting terahertz metamaterial, *Opt. Expr.*, 18:17504–17509, 2010. DOI: 10.1364/oe.18.017504. 5

[133] T. Driscoll, S. Palit, M. M. Qazilbash, M. Brehm, F. Keilmann, B. Chae, S. Yun, H. Kim, S. Cho, and N. M. Jokerst, Dynamic tuning of an infrared hybrid-metamaterial resonance using vanadium dioxide, *Appl. Phys. Lett.*, 93:24101, 2008. DOI: 10.1063/1.2956675. 5

[134] J. Gu, R. Singh, Z. Tian, W. Cao, Q. Xing, M. He, J. W. Zhang, J. Han, H. Chen, and W. Zhang, Terahertz superconductor metamaterial, *Appl. Phys. Lett.*, 97:71102, 2010. DOI: 10.1063/1.3479909. 5

[135] L. Ding, X. Luo, L. Cheng, M. Thway, J. Song, S. J. Chua, E. E. Chia, and J. Teng, Electrically and thermally tunable smooth silicon metasurfaces for broadband terahertz antireflection, *Adv. Opt. Mater.*, 6:1800928, 2018. DOI: 10.1002/adom.201800928. 5

[136] X. Yang, J. Yang, X. Hu, Y. Zhu, H. Yang, and Q. Gong, Multilayer-WS2: Ferroelectric composite for ultrafast tunable metamaterial-induced transparency applications, *Appl. Phys. Lett.*, 107:81110, 2015. DOI: 10.1063/1.4929701. 5

[137] X. Duan, S. Kamin, and N. Liu, Dynamic plasmonic colour display, *Nat. Commun.*, 8:14606, 2017. DOI: 10.1038/ncomms14606. 5

[138] S. Tang, T. Cai, H. X. Xu, Q. He, S. Sun, and L. Zhou, Multifunctional metasurfaces based on the merging, concept and anisotropic single-structure meta-atoms, *Appl. Sci.*, 8:555, 2018. DOI: 10.3390/app8040555. 7

[139] C. Zhang, F. Yue, D. Wen, M. Chen, Z. Zhang, W. Wang, and X. Chen, Multichannel metasurface for simultaneous control of holograms and twisted light beams, *ACS Photon.*, 4:1906–1912, 2017. DOI: 10.1021/acsphotonics.7b00587. 7

[140] T. Cai, G. M. Wang, H. X. Xu, S. W. Tang, H. Li, J. G. Liang, and Y. Q. Zhuang, Bifunctional Pancharatnam-Berry metasurface with high-efficiency helicity-dependent transmissions and reflections, *Ann. Phys.-Berlin*, 530:1700321, 2018. DOI: 10.1002/andp.201700321. 7, 78

[141] D. Wen, F. Yue, G. Li, G. Zheng, K. Chan, S. Chen, M. Chen, K. F. Li, P. W. H. Wong, and K. W. Cheah, Helicity multiplexed broadband metasurface holograms, *Nat. Commun.*, 6:8241, 2015. DOI: 10.1038/ncomms9241. 8, 12

[142] X. Ni, A. V. Kildishev, and V. M. Shalaev, Metasurface holograms for visible light, *Nat. Commun.*, 4:2807, 2013. DOI: 10.1038/ncomms3807. 9, 10

[143] Z. Zhang, D. Wen, C. Zhang, M. Chen, W. Wang, S. Chen, and X. Chen, Multifunctional light sword metasurface lens, *ACS Photon.*, 5:1794–1799, 2018. DOI: 10.1021/acsphotonics.7b01536. 10

[144] D. Wen, F. Yue, M. Ardron, and X. Chen, Multifunctional metasurface lens for imaging and Fourier transform, *Sci. Rep.*, 6:27628, 2016. DOI: 10.1038/srep27628. 10

[145] Y. Chen, X. Yang, and J. Gao, Spin-controlled wavefront shaping with plasmonic chiral geometric metasurfaces, *Light Sci. Applic.*, 7:84, 2018. DOI: 10.1038/s41377-018-0086-x. 10

[146] F. Zhang, M. Pu, X. Li, P. Gao, X. Ma, J. Luo, H. Yu, and X. Luo, All-dielectric meta-surfaces for simultaneous giant circular asymmetric transmission and wavefront shaping based on asymmetric photonic spin-orbit interactions, *Adv. Funct. Mater.*, 27:1704295, 2017. DOI: 10.1002/adfm.201704295. 12, 151

[147] P. Nayeri, F. Yang, and A. Z. Elsherbeni, Design and experiment of a single-feed quad-beam reflectarray antenna, *IEEE T. Antenn. Propag.*, 60:1166–1171, 2011. DOI: 10.1109/tap.2011.2173126. 19

[148] H. X. Xu, S. Tang, S. Ma, W. Luo, T. Cai, S. Sun, Q. He, and L. Zhou, Tunable microwave metasurfaces for high-performance operations: Dispersion compensation and dynamical switch, *Sci. Rep.*, 6:38255, 2016. DOI: 10.1038/srep38255. 24, 93

[149] S. Liu, T. J. Cui, Q. Xu, D. Bao, L. Du, X. Wan, W. X. Tang, C. Ouyang, X. Y. Zhou, and H. Yuan, Anisotropic coding metamaterials and their powerful manipulation of differently polarized terahertz waves, *Light Sci. Applic.*, 5:e16076, 2016. DOI: 10.1038/lsa.2016.76. 33

[150] T. Cai, G. Wang, X. Zhang, J. Liang, Y. Zhuang, D. Liu, and H. X. Xu, Ultra-thin polarization beam splitter using 2D transmissive phase gradient metasurface, *IEEE T. Antenn. Propag.*, 63:5629–5636, 2015. DOI: 10.1109/tap.2015.2496115. 33

[151] I. Yulevich, E. Maguid, N. Shitrit, D. Veksler, V. Kleiner, and E. Hasman, Optical mode control by geometric phase in quasicrystal metasurface, *Phys. Rev. Lett.*, 115:205501, 2015. DOI: 10.1103/physrevlett.115.205501. 40

[152] S. L. Jia, X. Wan, D. Bao, Y. J. Zhao, and T. J. Cui, Independent controls of orthogonally polarized transmitted waves using a Huygens metasurface, *Laser Photon. Rev.*, 9:545–553, 2015. DOI: 10.1002/lpor.201500094. 40

[153] K. Chen, Y. Feng, F. Monticone, J. Zhao, B. Zhu, T. Jiang, L. Zhang, Y. Kim, X. Ding, and S. Zhang, A reconfigurable active huygens' metalens, *Adv. Mater.*, 29:1606422, 2017. DOI: 10.1002/adma.201606422. 40

[154] S. Pancharatnam, Generalized theory of interference and its applications, *Proc. of the Indian Academy of Sciences*, 44:247–262, 1956. DOI: 10.1007/bf03046050. 59, 61

[155] M. V. Berry, Quantal phase factors accompanying adiabatic changes, *Proc. of the Royal Society of London. A. Mathematical and Physical Sciences*, 392:45–57, 1984. DOI: 10.1142/9789813221215_0006. 59, 61

[156] Z. E. Bomzon, G. Biener, V. Kleiner, and E. Hasman, Space-variant Pancharatnam-Berry phase optical elements with computer-generated subwavelength gratings, *Opt. Lett.*, 27:1141–1143, 2002. DOI: 10.1364/ol.27.001141. 61, 145

[157] N. Shitrit, I. Bretner, Y. Gorodetski, V. Kleiner, and E. Hasman, Optical spin hall effects in plasmonic chains, *Nano Lett.*, 11:2038–2042, 2011. DOI: 10.1021/nl2004835. 61, 62

[158] L. Huang, X. Chen, B. Bai, Q. Tan, G. Jin, T. Zentgraf, and S. Zhang, Helicity dependent directional surface plasmon polariton excitation using a metasurface with interfacial phase discontinuity, *Light Sci. Applic.*, 2:e70, 2013. DOI: 10.1038/lsa.2013.26. 61, 62, 68

[159] J. Lin, J. B. Mueller, Q. Wang, G. Yuan, N. Antoniou, X. Yuan, and F. Capasso, Polarization-controlled tunable directional coupling of surface plasmon polaritons, *Science*, 340:331–334, 2013. DOI: 10.1126/science.1233746. 61, 62, 68

[160] N. Shitrit, I. Yulevich, E. Maguid, D. Ozeri, D. Veksler, V. Kleiner, and E. Hasman, Spin-optical metamaterial route to spin-controlled photonics, *Science*, 340:724–726, 2013. DOI: 10.1126/science.1234892. 61, 62

[161] X. Chen, L. Huang, H. Mühlenbernd, G. Li, B. Bai, Q. Tan, G. Jin, C. Qiu, S. Zhang, and T. Zentgraf, Dual-polarity plasmonic metalens for visible light, *Nat. Commun.*, 3:1198, 2012. DOI: 10.1038/ncomms2207. 61, 138

[162] M. Kang, T. Feng, H. Wang, and J. Li, Wave front engineering from an array of thin aperture antennas, *Opt. Express*, 20:15882–15890, 2012. DOI: 10.1364/oe.20.015882. 61

[163] D. Sievenpiper, L. Zhang, R. F. Broas, N. G. Alexopolous, and E. Yablonovitch, High-impedance electromagnetic surfaces with a forbidden frequency band, *IEEE T. Microw. Theory*, 47:2059–2074, 1999. DOI: 10.1109/22.798001. 63

[164] J. M. Hao, L. Zhou, and C. T. Chan, An effective-medium model for high-impedance surfaces, *Appl. Phys. A*, 87:281–284, 2007. DOI: 10.1007/s00339-006-3825-4. 63

[165] J. Duan, H. Guo, S. Dong, T. Cai, W. Luo, Z. Liang, Q. He, L. Zhou, and S. Sun, High-efficiency chirality-modulated spoof surface plasmon meta-coupler, *Sci. Rep.*, 7:1354, 2017. DOI: 10.1038/s41598-017-01664-w. 69

[166] S. Sun, Q. He, J. Hao, S. Xiao, and L. Zhou, Electromagnetic metasurfaces: Physics and applications, *Adv. Opt. Photon.*, 11:380–479, 2019. DOI: 10.1364/AOP.11.000380. 70, 94

[167] Q. He, S. Sun, S. Xiao, and L. Zhou, High-efficiency metasurfaces: Principles, realizations, and applications, *Adv. Opt. Mater.*, 6:1800415, 2018. DOI: 10.1002/adom.201800415. 70, 94

[168] S. Jiang, X. Xiong, Y. Hu, S. Jiang, D. Xu, R. Peng, and M. Wang, High-efficiency generation of circularly polarized light via symmetry-induced anomalous reflection, *Phys. Rev. B*, 91:125421, 2015. DOI: 10.1103/physrevb.91.125421. 70

[169] M. Tymchenko, J. S. Gomez-Diaz, J. Lee, N. Nookala, M. A. Belkin, and A. Alù, Gradient nonlinear Pancharatnam-Berry metasurfaces, *Phys. Rev. Lett.*, 115:207403, 2015. DOI: 10.1103/physrevlett.115.207403. 70

[170] Y. Li, J. Zhang, S. Qu, J. Wang, H. Chen, L. Zheng, Z. Xu, and A. Zhang, Achieving wideband polarization-independent anomalous reflection for linearly polarized waves with dispersionless phase gradient metasurfaces, *J. Phys. D: Appl. Phys.*, 47:425103, 2014. DOI: 10.1088/0022-3727/47/42/425103. 70

[171] A. Arbabi and A. Faraon, Fundamental limits of ultrathin metasurfaces, *Sci. Rep.*, 7:43722, 2017. DOI: 10.1038/srep43722. 71

[172] X. Ding, F. Monticone, K. Zhang, L. Zhang, D. Gao, S. N. Burokur, A. de Lustrac, Q. Wu, C. W. Qiu, and A. Alù, Ultrathin Pancharatnam-Berry metasurface with maximal cross-polarization efficiency, *Adv. Mater.*, 27:1195–1200, 2015. DOI: 10.1002/adma.201405047. 71, 85

[173] W. Luo, S. Sun, H. X. Xu, Q. He, and L. Zhou, Transmissive ultrathin Pancharatnam-Berry metasurfaces with nearly 100% efficiency, *Phys. Rev. Appl.*, 7:44033, 2017. DOI: 10.1103/physrevapplied.7.044033. 71, 72, 75, 78, 79

[174] F. Monticone, N. M. Estakhri, and A. Alu, Full control of nanoscale optical transmission with a composite metascreen, *Phys. Rev. Lett.*, 110:203903, 2013. DOI: 10.1103/physrevlett.110.203903. 72

[175] M. Chen, M. Kim, A. M. Wong, and G. V. Eleftheriades, Huygens' metasurfaces from microwaves to optics: A review, *Nanophotonics*, 7:1207–1231, 2018. DOI: 10.1515/nanoph-2017-0117. 72

[176] C. Pfeiffer and A. Grbic, Controlling vector Bessel beams with metasurfaces, *Phys. Rev. Appl.*, 2:44012, 2014. DOI: 10.1103/physrevapplied.2.044012. 72

[177] S. Kruk, B. Hopkins, I. I. Kravchenko, A. Miroshnichenko, D. N. Neshev, and Y. S. Kivshar, Invited article: Broadband highly efficient dielectric metadevices for polarization control, *Apl. Photon.*, 1:30801, 2016. DOI: 10.1063/1.4949007. 72

[178] L. Wang, S. Kruk, H. Tang, T. Li, I. Kravchenko, D. N. Neshev, and Y. S. Kivshar, Grayscale transparent metasurface holograms, *Optica*, 3:1504–1505, 2016. DOI: 10.1364/optica.3.001504. 72

[179] L. Zhou, W. Wen, C. T. Chan, and P. Sheng, Electromagnetic-wave tunneling through negative-permittivity media with high magnetic fields, *Phys. Rev. Lett.*, 94:243905, 2005. DOI: 10.1103/physrevlett.94.243905. 72

[180] Z. Wang, S. Dong, W. Luo, M. Jia, Z. Liang, Q. He, S. Sun, and L. Zhou, High-efficiency generation of Bessel beams with transmissive metasurfaces, *Appl. Phys. Lett.*, 112:191901, 2018. DOI: 10.1063/1.5023553. 75, 78

[181] M. Jia, Z. Wang, H. Li, X. Wang, W. Luo, S. Sun, Y. Zhang, Q. He, and L. Zhou, Efficient manipulations of circularly polarized terahertz waves with transmissive meta-surfaces, *Light Sci. Applic.*, 8:16, 2019. DOI: 10.1038/s41377-019-0127-0. 76

[182] Z. Wu, X. Wang, W. Sun, S. Feng, P. Han, J. Ye, Y. Yu, and Y. Zhang, Vectorial diffraction properties of THz vortex Bessel beams, *Opt. Expr.*, 26:1506–1520, 2018. DOI: 10.1364/oe.26.001506. 78

[183] Z. Wu, X. Wang, W. Sun, S. Feng, P. Han, J. Ye, and Y. Zhang, Vector characterization of zero-order terahertz Bessel beams with linear and circular polarizations, *Sci. Rep.*, 7:13929, 2017. DOI: 10.1038/s41598-017-12524-y. 78

[184] H. X. Xu, S. Ma, W. Luo, T. Cai, S. Sun, Q. He, and L. Zhou, Aberration-free and functionality-switchable meta-lenses based on tunable metasurfaces, *Appl. Phys. Lett.*, 109:193506, 2016. DOI: 10.1063/1.4967438. 93

[185] R. Guzmán-Quirós, J. L. Gómez-Tornero, A. R. Weily, and Y. J. Guo, Electronically steerable 1D Fabry-Perot leaky-wave antenna employing a tunable high impedance surface, *IEEE T. Antenn. Propag.*, 60:5046–5055, 2012. DOI: 10.1109/tap.2012.2208089. 93

[186] J. Lee, S. Jung, P. Y. Chen, F. Lu, F. Demmerle, G. Boehm, M. C. Amann, A. Alù, and M. A. Belkin, Ultrafast electrically tunable polaritonic metasurfaces, *Adv. Opt. Mater.*, 2:1057–1063, 2014. DOI: 10.1002/adom.201400185. 93

[187] F. Ma, Y. Lin, X. Zhang, and C. Lee, Tunable multiband terahertz metamaterials using a reconfigurable electric split-ring resonator array, *Light Sci. Applic.*, 3:e171, 2014. DOI: 10.1038/lsa.2014.52. 93

[188] T. Jiang, Z. Wang, D. Li, J. Pan, B. Zhang, J. Huangfu, Y. Salamin, C. Li, and L. Ran, Low-DC voltage-controlled steering-antenna radome utilizing tunable active metamaterial, *IEEE T. Microw. Theory*, 60:170–178, 2011. DOI: 10.1109/tmtt.2011.2171981. 93

[189] T. H. Hand and S. A. Cummer, Reconfigurable reflectarray using addressable metamaterials, *IEEE Antenn. Wirel. Pr.*, 9:70–74, 2010. DOI: 10.1109/lawp.2010.2043211. 93

[190] P. Chen, J. Soric, Y. R. Padooru, H. M. Bernety, A. B. Yakovlev, and A. Alù, Nanostructured graphene metasurface for tunable terahertz cloaking, *New J. Phys.*, 15:123029, 2013. DOI: 10.1088/1367-2630/15/12/123029. 93

[191] H. Wakatsuchi, S. Kim, J. J. Rushton, and D. F. Sievenpiper, Waveform-dependent absorbing metasurfaces, *Phys. Rev. Lett.*, 111:245501, 2013. DOI: 10.1103/physrevlett.111.245501. 93

[192] P. Chen, C. Argyropoulos, and A. Alù, Broadening the cloaking bandwidth with non-foster metasurfaces, *Phys. Rev. Lett.*, 111:233001, 2013. DOI: 10.1103/physrevlett.111.233001. 93

[193] I. V. Shadrivov, P. V. Kapitanova, S. I. Maslovski, and Y. S. Kivshar, Metamaterials controlled with light, *Phys. Rev. Lett.*, 109:83902, 2012. DOI: 10.1103/physrevlett.109.083902. 93

[194] N. Kaina, M. Dupré, G. Lerosey, and M. Fink, Shaping complex microwave fields in reverberating media with binary tunable metasurfaces, *Sci. Rep.*, 4:6693, 2014. DOI: 10.1038/srep06693. 93

[195] T. J. Cui, M. Q. Qi, X. Wan, J. Zhao, and Q. Cheng, Coding metamaterials, digital metamaterials and programmable metamaterials, *Light Sci. Applic.*, 3:e218, 2014. DOI: 10.1038/lsa.2014.99. 93

[196] B. O. Zhu, K. Chen, N. Jia, L. Sun, J. Zhao, T. Jiang, and Y. Feng, Dynamic control of electromagnetic wave propagation with the equivalent principle inspired tunable metasurface, *Sci. Rep.*, 4:4971, 2014. DOI: 10.1038/srep04971. 93

[197] H. X. Xu, S. Ma, X. Ling, X. Zhang, S. Tang, T. Cai, S. Sun, Q. He, and L. Zhou, Deterministic approach to achieve broadband polarization-independent diffusive scatterings based on metasurfaces, *ACS Photon.*, 5:1691–1702, 2017. DOI: 10.1021/acsphotonics.7b01036. 94

[198] H. X. Xu, L. Han, Y. Li, Y. Sun, J. Zhao, S. Zhang, and C. Qiu, Completely spin-decoupled dual-phase hybrid metasurfaces for arbitrary wavefront control, *ACS Photon.*, 6:211–220, 2018. DOI: 10.1021/acsphotonics.8b01439. 94

[199] Q. He, S. Sun, and L. Zhou, Tunable/reconfigurable metasurfaces: Physics and applications, *Research*, 2019:1849272, 2019. DOI: 10.34133/2019/1849272. 94

[200] H. X. Xu, G. Hu, Y. Li, L. Han, J. Zhao, Y. Sun, F. Yuan, G. Wang, Z. H. Jiang, and X. Ling, Interference-assisted kaleidoscopic meta-plexer for arbitrary spin-wavefront manipulation, *Light Sci. Applic.*, 8:3, 2019. DOI: 10.1038/s41377-018-0113-y. 94

[201] S. V. Hum, M. Okoniewski, and R. J. Davies, Modeling and design of electronically tunable reflectarrays, *IEEE T. Antenn. Propag.*, 55:2200–2210, 2007. DOI: 10.1109/tap.2007.902002. 95

[202] M. Riel and J. Laurin, Design of an electronically beam scanning reflectarray using aperture-coupled elements, *IEEE T. Antenn. Propag.*, 55:1260–1266, 2007. DOI: 10.1109/tap.2007.895586. 95

[203] K. K. Kishor and S. V. Hum, An amplifying reconfigurable reflectarray antenna, *IEEE T. Antenn. Propag.*, 60:197–205, 2011. DOI: 10.1109/tap.2011.2167939. 95

[204] M. Sazegar, Y. Zheng, C. Kohler, H. Maune, M. Nikfalazar, J. R. Binder, and R. Jakoby, Beam steering transmitarray using tunable frequency selective surface with integrated ferroelectric varactors, *IEEE T. Antenn. Propag.*, 60:5690–5699, 2012. DOI: 10.1109/tap.2012.2213057. 95

[205] A. Clemente, L. Dussopt, R. Sauleau, P. Potier, and P. Pouliguen, Wideband 400-element electronically reconfigurable transmitarray in X band, *IEEE T. Antenn. Propag.*, 61:5017–5027, 2013. DOI: 10.1109/tap.2013.2271493. 95, 121

[206] Smv1405-Smv1430 series: Plastic packaged abrupt junction tuning varactors, 2013. (Date of access: 10/05/2015.) 98

[207] f_2 is also slightly tuned. However, since the varactor diode is connected the central bar of the "H", structure, it affects f_1 more significantly than on f_2. 115

[208] H. X. Xu, G. Wang, T. Cai, J. Xiao, and Y. Zhuang, Tunable Pancharatnam-Berry metasurface for dynamical and high-efficiency anomalous reflection, *Opt. Expr.*, 24:27836–27848, 2016. DOI: 10.1364/oe.24.027836. 123, 138, 145

[209] C. Qu, S. Ma, J. Hao, M. Qiu, X. Li, S. Xiao, Z. Miao, N. Dai, Q. He, and S. Sun, Tailor the functionalities of metasurfaces based on a complete phase diagram, *Phys. Rev. Lett.*, 115:235503, 2015. DOI: 10.1103/physrevlett.115.235503. 124

[210] Smp1345 series: Very low capacitance, plastic packaged silicon pin diodes, 2012. (Date of access: 10/05/2015.) 126

[211] H. Haus, *Waves and Fields in Optoelectronics*, Prentice Hall, Englewood Cliffs, NJ, 1984. 134

[212] S. Fan, W. Suh, and J. D. Joannopoulos, Temporal coupled-mode theory for the Fano resonance in optical resonators, *J. Opt. Soc. Amer. A*, 20:569–572, 2003. DOI: 10.1364/josaa.20.000569. 134

[213] W. Suh, Z. Wang, and S. Fan, Temporal coupled-mode theory and the presence of non-orthogonal modes in lossless multimode cavities, *IEEE J. Quantum Elect.*, 40:1511–1518, 2004. DOI: 10.1109/jqe.2004.834773. 134

[214] L. Zhou, H. Li, Y. Qin, Z. Wei, and C. T. Chan, Directive emissions from sub-wavelength metamaterial-based cavities, *Appl. Phys. Lett.*, 86:101101, 2005. DOI: 10.1063/1.1881797. 136

[215] P. Genevet, N. Yu, F. Aieta, J. Lin, M. A. Kats, R. Blanchard, M. O. Scully, Z. Gaburro, and F. Capasso, Ultra-thin plasmonic optical vortex plate based on phase discontinuities, *Appl. Phys. Lett.*, 100:13101, 2012. DOI: 10.1063/1.3673334. 137

[216] E. Karimi, S. A. Schulz, I. De Leon, H. Qassim, J. Upham, and R. W. Boyd, Generating optical orbital angular momentum at visible wavelengths using a plasmonic metasurface, *Light Sci. Applic.*, 3:e167, 2014. DOI: 10.1038/lsa.2014.48. 138

[217] M. Zhou, S. B. Sørensen, O. S. Kim, E. Jørgensen, P. Meincke, O. Breinbjerg, and G. Toso, The generalized direct optimization technique for printed reflectarrays, *IEEE T. Antenn. Propag.*, 62:1690–1700, 2013. DOI: 10.1109/tap.2013.2254446. 138

[218] M. Zhou, S. B. Sørensen, O. S. Kim, E. Jørgensen, P. Meincke, and O. Breinbjerg, Direct optimization of printed reflectarrays for contoured beam satellite antenna applications, *IEEE T. Antenn. Propag.*, 61:1995–2004, 2012. DOI: 10.1109/tap.2012.2232037. 138

[219] E. Erdil, K. Topalli, N. S. Esmaeilzad, Ö. Zorlu, H. Kulah, and O. A. Civi, Re-configurable nested ring-split ring transmitarray unit cell employing the element rotation method by microfluidics, *IEEE T. Antenn. Propag.*, 63:1163–1167, 2015. DOI: 10.1109/tap.2014.2387424. 138

[220] M. Euler and V. F. Fusco, Frequency selective surface using nested split ring slot elements as a lens with mechanically reconfigurable beam steering capability, *IEEE T. Antenn. Propag.*, 58:3417–3421, 2010. DOI: 10.1109/tap.2010.2055814. 138

[221] J. Huang and R. J. Pogorzelski, A Ka-band microstrip reflectarray with elements having variable rotation angles, *IEEE T. Antenn. Propag.*, 46:650–656, 1998. DOI: 10.1109/8.668907. 138

[222] A. E. Martynyuk, J. M. Lopez, and N. A. Martynyuk, Spiraphase-type reflectarrays based on loaded ring slot resonators, *IEEE T. Antenn. Propag.*, 52:142–153, 2004. DOI: 10.1109/tap.2003.820976. 138

[223] B. Strassner, C. Han, and K. Chang, Circularly polarized reflectarray with microstrip ring elements having variable rotation angles, *IEEE T. Antenn. Propag.*, 52:1122–1125, 2004. DOI: 10.1109/tap.2004.825635. 138

[224] R. H. Phillion and M. Okoniewski, Lenses for circular polarization using planar arrays of rotated passive elements, *IEEE T. Antenn. Propag.*, 59:1217–1227, 2011. DOI: 10.1109/tap.2011.2109694. 138, 144

[225] A. Yu, F. Yang, A. Z. Elsherbeni, J. Huang, and Y. Kim, An offset-fed X-band reflectarray antenna using a modified element rotation technique, *IEEE T. Antenn. Propag.*, 60:1619–1624, 2011. DOI: 10.1109/tap.2011.2180299. 138

[226] J. Rodriguez-Zamudio, J. I. Martinez-Lopez, J. Rodriguez-Cuevas, and A. E. Martynyuk, Reconfigurable reflectarrays based on optimized spiraphase-type elements, *IEEE T. Antenn. Propag.*, 60:1821–1830, 2012. DOI: 10.1109/tap.2012.2186231. 138

[227] MA4PBl027, HMICTM Silicon Beam-Lead PIN Diodes, www.macomtech.com for data sheets and product information, M/A-COM Technology Solutions, Inc. 140, 141

[228] A. Komar, R. Paniagua-Dominguez, A. Miroshnichenko, Y. F. Yu, Y. S. Kivshar, A. I. Kuznetsov, and D. Neshev, Dynamic beam switching by liquid crystal tunable dielectric metasurfaces, *ACS Photon.*, 5:1742–1748, 2018. DOI: 10.1021/acsphotonics.7b01343. 151

[229] K. Chen, G. Ding, G. Hu, Z. Jin, J. Zhao, Y. Feng, T. Jiang, A. Alù, and C. W. Qiu, Directional janus metasurface, *Adv. Mater.*, 1906352, 2019. DOI: 10.1002/adma.201906352. 152

[230] W. Pan, T. Cai, S. Tang, L. Zhou, and J. Dong, Trifunctional metasurfaces: Concept and characterizations, *Opt. Expr.*, 26:17447–17457, 2018. DOI: 10.1364/oe.26.017447. 152

[231] A. Shaltout, A. Kildishev, and V. Shalaev, Time-varying metasurfaces and Lorentz nonreciprocity, *Opt. Mater. Expr.*, 5:2459–2467, 2015. DOI: 10.1364/ome.5.002459. 152

[232] J. Y. Dai, J. Zhao, Q. Cheng, and T. J. Cui, Independent control of harmonic amplitudes and phases via a time-domain digital coding metasurface, *Light Sci. Applic.*, 7:90, 2018. DOI: 10.1038/s41377-018-0092-z. 152

[233] L. Zhang, X. Q. Chen, S. Liu, Q. Zhang, J. Zhao, J. Y. Dai, G. D. Bai, X. Wan, Q. Cheng, and G. Castaldi, Space-time-coding digital metasurfaces, *Nat. Commun.*, 9:4334, 2018. DOI: 10.1038/s41467-018-06802-0. 152

[234] B. Assouar, B. Liang, Y. Wu, Y. Li, J. Cheng, and Y. Jing, Acoustic metasurfaces, *Nat. Rev. Mater.*, 3:460–472, 2018. DOI: 10.1038/s41578-018-0061-4. 152

[235] Y. Li, C. Shen, Y. Xie, J. Li, W. Wang, S. A. Cummer, and Y. Jing, Tunable asymmetric transmission via lossy acoustic metasurfaces, *Phys. Rev. Lett.*, 119:35501, 2017. DOI: 10.1103/physrevlett.119.035501. 152

[236] B. Xie, K. Tang, H. Cheng, Z. Liu, S. Chen, and J. Tian, Coding acoustic metasurfaces, *Adv. Mater.*, 29:1603507, 2017. DOI: 10.1002/adma.201603507. 152

[237] S. A. Cummer, J. Christensen, and A. Alù, Controlling sound with acoustic metamaterials, *Nat. Rev. Mater.*, 1:16001, 2016. DOI: 10.1038/natrevmats.2016.1. 152

[238] S. Zou, Y. Xu, R. Zatianina, C. Li, X. Liang, L. Zhu, Y. Zhang, G. Liu, Q. H. Liu, and H. Chen, Broadband waveguide cloak for water waves, *Phys. Rev. Lett.*, 123:74501, 2019. DOI: 10.1103/physrevlett.123.074501. 152

[239] Y. Li, X. Shen, Z. Wu, J. Huang, Y. Chen, Y. Ni, and J. Huang, Temperature-dependent transformation thermotics: From switchable thermal cloaks to macroscopic thermal diodes, *Phys. Rev. Lett.*, 115:195503, 2015. DOI: 10.1103/physrevlett.115.195503. 152

[240] D. M. Nguyen, H. Xu, Y. Zhang, and B. Zhang, Active thermal cloak, *Appl. Phys. Lett.*, 107:121901, 2015. DOI: 10.1063/1.4930989. 152

Authors' Biographies

HE-XIU XU

He-Xiu Xu received his Ph.D. in Electronic Science and Technology from the Air Force Engineering University, China, in 2014. From 2015–2017, he was a postdoctoral fellow of the Physics Department at Fudan University (Shanghai, China). In 2017–2018, he was a visiting scholar in the Department of Electrical and Computer Engineering of the National University of Singapore. He joined the Department of Electronic Science and Technology of Air Force Engineering University in 2014 as an assistant professor, became an associate professor in 2016, and is now a full professor since 2019. He has been working in the fields of metamaterials, metasurfaces, and their potential applications in circuits and functional devices, and has published more than 120 papers in scientific journals. He was elected as a fellow of The Institution of Engineering and Technology (IET) in 2019.

SHIWEI TANG

Shiwei Tang received his Ph.D. in the Physics Department of Fudan University, Shanghai, China, in 2014. He was a postdoctoral fellow in the Materials Science Department of Fudan University from 2014–2015. He joined Ningbo University, Ningbo, China in 2016 and was promoted to an Associate Professor in 2019. His current research interests include metamaterials/metasurfaces, microcavities, plasmonics, and nanophotonics. He has published over 60 papers in journals such as *Advanced Materials*, *Advanced Functional Materials*, *ACS Nano*, and *Optics Express*.

TONG CAI

Tong Cai received the B.S. and Ph.D. in Electrical Engineering from the Air Force Engineering University, Xi'an, China, in 2012, and 2017, respectively. He was with Fudan University as a visiting scholar from 2015–2017. He was with the Air Force Engineering University, where he became a Lecturer in 2017 and an associate professor in 2020, and has been a Post-Doctoral Researcher with Zhejiang University since 2019. His research interests include metamaterials, metasurfaces, and their applications to novel antennas and multifunctional devices. He obtained the support of the Postdoctoral Innovation Talents Support Program of China in 2019. He has authored over 40 peer-reviewed first author articles in journals such as *Advanced Photonics*, *Ad-*

vanced Optical Materials, IEEE Transactions on Antennas and Propagations, and *Physical Review Applied*.

SHULIN SUN

Shulin Sun received his Ph.D. in Physics at Fudan University in 2009. From 2010–2013, he was a Postdoctoral Fellow of the Department of Physics at National Taiwan University. In 2013, he joined the Department of Optical Science and Engineering at Fudan University, and has been a full Professor and associate dean of the department since 2019. He has been working in the fields of metamaterials/metasurfaces, plasmonics, and photonic crystals, and published over 70 papers in journals such as *Nature Materials, Nano Letters, Advances in Optics and Photonics*, and *Light: Science & Applications*.

QIONG HE

Qiong He received his Ph.D. degree in Physics from the Paris Institute of Optics in Paris-Sud University (Orsay, France) in 2008. From 2009–2013, he was a postdoctoral fellow in the Physics Department of Fudan University. He is currently an associate professor in the Physics Department of Fudan University (Shanghai, China). His research interests focus on metamaterials and plasmonics. He has coauthored more than 80 publications in scientific journals, including *Nature Materials, Physics Review X, Physics Review Letters, Advances in Optics and Photonics, Light: Science & Applications*, and *Nano Letters*.

LEI ZHOU

Lei Zhou received his Ph.D. in Physics from Fudan University, China, in 1997. From 1997–2000, he was a postdoctoral fellow of the Institute for Material Research at Tohoku University (Sendai, Japan). In 2000–2004, he was a visiting scholar in the Physics Department of the Hong Kong University of Science and Technology. He joined the Physics Department of Fudan University in 2004 as a professor, became a "Xi-De" chair professor in 2013, and is now Chair of the department. He has been working in the fields of magnetism, metamaterials, photonic crystals, and plasmonics, and has published more than 180 papers in scientific journals. He was elected as a fellow of The Optical Society (OSA) in 2019, and a Clarivate Analytics Global Highly Cited Researcher (2019–2020).